Printed especially for Contributing Members
of the Smithsonian National Associate Program

DRAWN FROM NATURE

SMITHSONIAN INSTITUTION PRESS, WASHINGTON, D.C., 1984

The Botanical Art of

DRAWN FROM NATURE

Joseph Prestele and His Sons

CHARLES VAN RAVENSWAAY

Printed and bound in Great Britain

Library of Congress Cataloging in Publication Data

van Ravenswaay, Charles.
Drawn from nature.
Includes bibliographical references and index.
Supt. of Docs no.: SI 1.2:N21/2
1. Prestele, Joseph, 1796–1867. 2. Prestele family. 3. Botanical artists—United States—Biography. 4. Botanical artists—Germany (West)—Biography. I. Title.
QK98.183.P74V36 1984 580'.22 84-600003
ISBN 0-87474-938-7 (alk. paper)

The paper in this book meets the guidelines for permanence and durability of the Committee on Production Guidelines for Book Longevity of the Council on Library Resources.

Contents

To Edgar P. Richardson

Acknowledgments

MY SEARCH FOR INFORMATION ABOUT THE PRESTELES WAS DETECTIVE WORK of a sort, more so than is usual in research projects. I began with only a name and several faint clues. One clue led to another and so the search expanded.

It soon became apparent that while seeking information about Joseph Prestele and his family, I had to learn about their times, their associates, and the religious community with which I found they were associated. Some knowledge about the history of lithography was also essential, particularly the obscure techniques developed early in the nineteenth century and almost forgotten today. References to the different qualities of lithographic stones, and to the necessity for artists to size the paper they used for prints, led to other studies. As my search, and my studies, grew, I was aided by the mysterious workings of serendipity, and particularly by many wonderfully helpful friends, both old and new. Their assistance is acknowledged with the greatest appreciation.

Among these helpful people were: Janet H. Baker, Media, Pennsylvania; Christopher P. Bickford, director, Connecticut Historical Society; Bernadette G. Callery, librarian, Hunt Institute for Botanical Documentation; Anne F. Clapp, while print and paper conservator, Henry Francis du Pont Winterthur Museum, and later in private practice; Lenore M. Dickinson, while librarian, Gray Herbarium of Harvard University; Joan Liffring-Zug, Iowa City; Marjorie B. Cohn, conservator of art on paper, Fogg Art Museum, Harvard; Laura Dulinawka, secretary, West Seneca Historical Society, Inc., West Seneca, New York; E. McSherry Fowble, assistant curator in charge of graphics, H. F. du Pont Winterthur Museum; Helen Ver Nooy Gearn, city historian, Newburgh, New York; Karen Laughlin, librarian, Iowa State Historical Society; William H. Loos, curator, Rare Book Room, Buffalo and Erie County Public Library, Buffalo, New York; Dorothy S. Manks, for many years librarian of the Massachusetts Horticultural Society; Barbara L. Mykrantz, archivist, Missouri Botanical Garden, St. Louis; Nathaniel

H. Puffer, assistant director of libraries, University of Delaware; John F. Reed, librarian, New York Botanical Garden; George J. Reilly, coordinator and head of scientific research, Henry Francis du Pont Winterthur Museum; Evald Rink, former head cataloguer, Eleutherian Mills Historical Library, Wilmington, Delaware; Frank H. Sommer III, head, Libraries Division, Henry Francis du Pont Winterthur Museum; and Mary Thomas, assistant librarian, Gray Herbarium of Harvard University, who was particularly helpful in locating obscure publications and expediting photography.

Also helpful were Ruth F. Schallert, botany branch librarian, Smithsonian Institution; Carole Taylor, West Seneca, New York; Judith A. Weinberg, librarian, Massachusetts Horticultural Society, Boston; and Elisabeth Woodburn, Hopewell, New Jersey, friend, scholar, and bookseller who first directed my attention to Joseph Prestele.

Greg Koos, archivist, McLean County Historical Society, Bloomington, Illinois, with unflagging enthusiasm, uncovered information that gave substance to an important section of the manuscript. Robert A. Tibbetts, curator of Special Collections at the Ohio State University Libraries, sent me detailed information about their Gottlieb Prestele nurserymen's plates. David W. Corson, history of science librarian, Cornell University Libraries, and Susan Markowitz, reference librarian, Mann Library, also at Cornell, supplied data about the Prestele plates in their collections. Details about the work of William Henry Prestele at the Department of Agriculture came from J. T. McGrew, research plant pathologist, Beltsville Agricultural Research Center, Maryland.

During the early period of my research, Mrs. Ferdinand Ruff, a volunteer assistant at the Museum of Amana History, aided me with data from the museum's collection. Later, in the summer of 1982, I was fortunate in obtaining the assistance of Mary Fredericksen, then a graduate student at the University of Iowa, to contact descendants of the Inspirationists living in the Amanas for additional information. During her canvass she was aided by many people whose assistance then, and in 1983 when I visited Amana, has been of the greatest help. They include Elmer P. Graesser, Mrs. Madeline Roemig, Mrs. Adolph Schmieder, Mrs. Fred Schneider, Don Shoup, Mrs. Lina Unglenk, Mrs. Emma Setzer, and Mrs. Joseph Mattes, among others. Lanny Haldy, the director of the Museum of Amana History, assisted in various useful ways, sharing with me his sensitive understanding of the Amana heritage, and information relating to its factual history. Two long Prestele letters which he provided, translated by Mrs. Magdalina Schuerer, supplied important details about Joseph Prestele's immigration to America.

Through a chance meeting with Alma Burner Creek, of the Department of Rare Books and Special Collections, University of Rochester Libraries, I learned of important Prestele and Dewey material in the Ellwanger and Barry Collection under her charge. The four days spent exploring that rich collection was made both productive and pleasant by the unfailing courtesies of all the staff. Particular thanks are due Mrs. Creek, and Karl S. Kabelac, manuscripts librarian, who were most generous with their assistance both during and following my visit.

Permission to quote from the Torrey Papers at the New York Botanical Garden was given by the director of the library, Charles R. Long; and from the Asa Gray Papers in the Library of the Gray Herbarium of Harvard University, by the director, Otto T. Solbrig.

Throughout my work on the Prestele story, Elisabeth Kottenhahn of Wilmington has patiently supplied translations, and through her contacts in West Germany she enlisted the assistance of the late Frau Hermine Stickel, Stuttgart, and Herr Studienrat Theodor Schnürle, living near Ulm; also Dekan Horst Grimm, pastor of St. Martin's Roman Catholic Church, Jettingen, who provided data and art.

Other research assistance in West Germany came from Mrs. Richard E. Schroeder, formerly of Washington, D.C., and now a resident of Munich, who located in the Bavarian State Library examples of art produced by the youthful Joseph Prestele. These watercolors were copied through the kindness of Frau Liselotte Renner, Handscriften u. Inkunabelsammlung, of the Bavarian State Library.

Dr. D. G. Huttleston, taxonomist, Longwood Gardens, Kennett Square, Pennsylvania, very kindly reviewed the scientific names of the plants given on the plates.

The assistance of the Barra Foundation in the publication of this book, and the wise and helpful suggestions of the foundation's publication consultant, Regina Ryan, are acknowledged with gratitude.

The first draft of the manuscript was read by Dr. Edgar P. Richardson, Philadelphia, a generous and helpful act of friendship.

Preface

The Search for Joseph Prestele

DURING THE 1970S, DR. FRITZ H. HERRMANN, A SCHOLAR LIVING AT FRIEDberg, Hessen, West Germany, came upon a publication on the most poisonous plants of Germany, published in 1843. It consisted of two parts, printed separately, one containing a short text by Carl Soldan, identified as a schoolmaster; the other, a portfolio with twenty-four large, "absolutely enchanting" colored plates, as Dr. Herrmann described them. These engraved illustrations were by an artist named Joseph Prestele and were the remarkable production of a small print shop in Dr. Herrmann's own city of Friedberg. Both Dr. Herrmann and his wife, Dr. Lore Herrmann, were intrigued and sought to learn about the artist. In time they published their findings.[1]

A few years later, before I had ever heard of Joseph Prestele or of the Herrmann's studies, I had an experience like theirs. It happened when, by chance, I bought an odd Smithsonian publication titled *Plates Prepared Between The Years 1849 and 1859, to Accompany a Report on the Forest Trees of North America, by Asa Gray.* It was more of a portfolio than a book for it consisted of twenty-three lithographed and hand-colored illustrations,* prefaced by a two-page "Advertisement," or statement, by Dr. Samuel P. Langley, then secretary of the Smithsonian. Dr. Langley explained that the plates (full-page illustrations) were the only tangible result of a project conceived more than forty years earlier by Professor Joseph Henry, the first secretary of the Smithsonian Institution, in cooperation with his friend and colleague Dr. Asa Gray, professor of botany at Harvard.

Dr. Langley's statement seemed rather vague. Why, I wondered, was the work never published? The plates were striking. They were wonderfully delicate

---

* *A checklist of these plates is given in Appendix A.*

in line and subtle in coloring and obviously had been created by an artist, or artists, who knew and loved plants. Nearly half lacked signatures; most of the others showed that the drawings for them had been made by Isaac Sprague, a familiar name among American botanical artists of the nineteenth century. Others credited "J. Prestele" as having engraved—and in several instances having engraved *and* printed—them. But who was "J. Prestele"? And what was the story behind Langley's vague account of the book that was never published?

Mr. Langley provided few clues, but with them my search for Joseph Prestele began.

In time I learned of the interest of the Herrmanns in Prestele and obtained copies of their publications. Their search for answers had led in many directions, often with interesting results, as mine began to do. We learned that the career of this quiet and self-effacing artist touched people and events important in the history of nineteenth-century botany in Europe and America and in the development of lithography. We were moved by the personal story of the man, particularly his abiding religious faith and that of his gentle little wife Karolina, and their involvement in a communal religious sect whose search for a refuge eventually led them from Germany to the Iowa frontier.

Most important of all for me, was the realization that although the art of the Presteles in America had been forgotten for many years, the flowers, fruits, and trees they portrayed with such loving care will always give pleasure, for they recall the fresh beauty of America's springtime.

This book resulted from my search for Joseph Prestele and his sons.

CHARLES VAN RAVENSWAAY

List of Color Plates

These plates are found on pages 131 through 319.

PART I

Joseph Prestele, 1796–1867

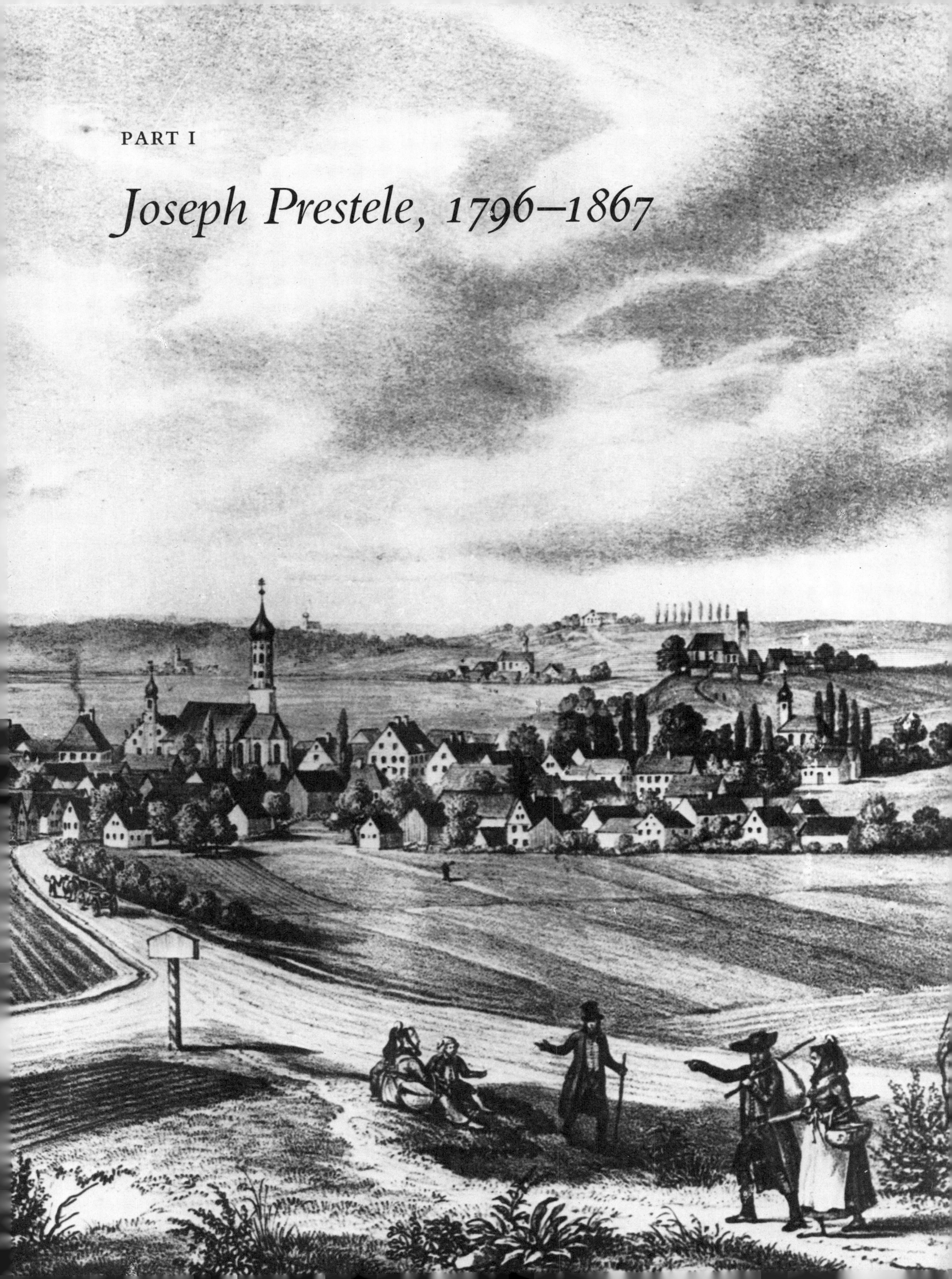

Jettingen, Upper Bavaria, the birthplace of Joseph Prestele, as it appeared about 1830.

Markt Jettingen, Sitz des Freiherrsch von Stauffenbergschen Patromonialgerichts 1. Klasse (Market town of Jettingen, seat of Baron von Stauffenberg's hereditary jurisdiction). Lithograph; drawn from nature by Nicolaus Lechner; drawn on stone by Gustav Kraus; ca. 1830; 44.5 × 50 cm. Courtesy, Dekan Horst Grimm, pastor of St. Martin's Roman Catholic Church, Jettingen, Bavaria, West Germany.

Gustav Wilhelm Kraus (1804–Munich 1852) was a painter of landscapes, buildings, and monuments.

1 The Bavarian Years, 1796–1843

Beginnings of a Botanical Artist

THE PROVINCE OF UPPER BAVARIA, WHOSE WESTERN BOUNDARY IS THE DANube, is a gentle countryside of rolling hills and clear streams, of woodlands, meadows, and neatly tended fields. Castles and old churches are reminders of past ages. It is a region which, for all its quiet beauty, has known too many armies during the centuries, and which knew the tensions of the Protestant Reformation.

In the countryside, somewhat away from the old road between Ulm on the Danube and Augsburg to the east, is Jettingen which during the late eighteenth century was much like other villages hereabouts. Near one end of the straggling community was the parish church of St. Martin with its slender baroque steeple, and the Rathaus, or Town Hall, alongside. At the other end of the village was Schloss Stauffenberg, a massive building with round towers at each of its four corners. Stark of line and mass it proclaimed authority. Adjoining the schloss were meadows, an orchard, and a garden. Lines of trees, many being slender poplars, marked boundaries between fields. Between church and schloss the homes and shops of the villagers lined a few twisting old streets. Close by them were farm meadows and croplands.

It was in this pastoral community that Joseph Prestele was born early in November 1796. He was baptized soon afterward—on November 8—in St. Martin's Roman Catholic Church, Jettingen. The entry in the baptismal register states that he was the son of Martin and Franziska Prestele, and was given the name of Joseph Martin.[1] In his young manhood he sometimes signed himself "Franz Joseph Prestele"; his army discharge of 1823 refers to him as "Martin Prestele." In his professional career he consistently used "Joseph Prestele."[2]

The artist's father is said to have been caretaker on the estate of a count—presumably Count von Stauffenberg. One would like to believe that young Joseph

Die 8va Novi

Infans Joseph Martinus
Parent. Martinus Proſtele, et Franciſca
Patr. D. Franciſcus Baugger oeconomio
Profectus, et Creſcentia Isenbühlinn,
Cujus Loco ſtetit M. anna Proſtelinn.

FIG. 1 Joseph Prestele's Record of Baptism.

Entry in the Register of Baptisms, St. Martin's Church, Jettingen, Upper Bavaria, recording the baptism of Joseph Martin Prostele [*sic*], 8 November 1796. The parents are given as Martin Prostele and Franziska (his wife); the godparents, D. Francis Baugger, farmer, and Crescentia Isenbuehl, in whose place stood M. Anna Prostele. Courtesy, Dekan Horst Grimm, pastor of St. Martin's Church.

first developed his lifelong interest in plants among the gardens, and orchards, and flower-bordered lanes of the count's estate, so charmingly recorded in an early nineteenth-century print.[3]

There is a story that after Joseph had completed his village schooling, he then had some instruction in botany at a nursery (*Gärtnerei*), where flowers, shrubs, trees, and sometimes vegetables were grown and sold. It seems he also studied drawing and painting flowers and fruits. Later he pursued his art studies in Vienna under Johann Knapp (1778–1833), who was one of the better flower painters of his time. Knapp's talents were recognized in 1804 by his appointment as painter to the Archduke Anton of Austria.[4] Katharine S. White in *Onward and Upward in the Garden* commented on his ability to portray realistically on canvas "florid but luscious bouquets."[5]

Prestele came to live in Munich where lithography had been invented about 1796, the year he was born, and which had become a center for the development of that process during the early nineteenth century. So it was that along with his training in drawing, painting, and botany, Prestele also became skilled in various lithographic techniques. Later he variously described himself as "Painter," "Painter and Lithographer," and "Engraver." Twentieth-century French and German dictionaries of artists refer to him as a "painter of flowers and lithographer."[6]

Prestele demonstrated his exceptional artistic talent very early. In 1812, when he was sixteen, he produced a manuscript portfolio of forty-eight watercolors of plants, carefully detailed, with the title *Icones plantarum selectarum Horti Regii Academici Monacensis vivis coloribus pictae a Fr. Jos. Prestele* (Illustrations of selected plants in the garden of the Royal Academy of Munich, portrayed in natural colors

ICONES
PLANTARUM
SELECTARUM
HORTI
REGII ACADEMICI
MONACENSIS
vivis coloribus picta
A
FR. JOS. PRESTELE.

FIG. 2 Title page of *Icones plantarum selectarum Horti Regii Academici Monacensis*, by Joseph Prestele, 1812, watercolor. Courtesy, the Bavarian State Library, Munich, West Germany, and reproduced here by permission.

FIG. 3 "*Strelitzia Reginae.*" (Queen's Bird of Paradise), by Joseph Prestele, watercolor. Plate 13 in *Icones plantarum selectarum*, 1812. Courtesy, the Bavarian State Library, Munich, West Germany, and reproduced here by permission.

FIG. 4 "*Ferraria Pavonia. Pfauenartige Ferrarie.*" (*Tigridia pavonia.* Common Tigerflower.) "Joseph Prestele pinx.," watercolor. Plate 32 in *Icones plantarum selectarum*, 1812. Courtesy, the Bavarian State Library, Munich, West Germany, and reproduced here by permission.

by Franz Joseph Prestele). In discussing "the large sized portfolio," the scholar Dr. Fritz H. Herrmann remarked that while the youthful exuberance of the artist is suggested by the fanciful title page with its elaborate border, its medley of colors, and the varied designs of lettering, the plants were painted with mature and painstaking care.[7] (See also Plates 1–3.)

Four years later, when he was twenty, he was employed at the Royal Botanical Garden in Munich—first, perhaps, as a gardener. His skill in drawing and painting must soon have been recognized for he was made a staff artist working mainly in watercolors (aquarelle). He is said to have drawn many of the plants for reproduction in the *Flora Monacensis* (Flowers of Munich, published 1811–18), which the artist Johann Nepomuk Mayrhofer copied on stone for lithographing, and for which the garden's director, Franz von Paula von Schrank, wrote the text.

In May 1818, Prestele began his required period of military service, during which he was a cannoneer in an artillery regiment for a period of five years and

FIG. 5 "*Passiflora Alata.*" (Wingstem Passiflora), by Joseph Prestele, watercolor. Plate 19 in *Icones plantarum selectarum*, 1812. Courtesy, the Bavarian State Library, Munich, West Germany, and reproduced here by permission.

FIG. 6 Graves of the Prestele Children. Unsigned watercolor; painted by Joseph Prestele, Sr.; 10 × 12.5 cm. Courtesy, The Amana Heritage Society, Museum of Amana History.

This small painting is of the graves of two of Prestele's daughters, Karolina Amalia Barbara (1825–1833) and Franziska Augusta (1830–1834). It was painted in (Munich?) Bavaria, about 1834.

four months. While in service he seems to have continued working at the Royal Botanical Garden, and he was involved in other non-military activities, which suggest that his military duties were not full time.

While he was in the army another collection of his lithographs of ornamental plants was published in 1819–20 in three parts, each with six pages, in royal folio (20 by 25 inches). This he humbly dedicated to Her Royal Highness, Friederike Wilhelmine Karoline, Queen of Bavaria. In 1820 also appeared his *Anfangsgründe zur Blumenzeichnung* (Basic instructions for drawing flowers), whose plates were described by a reviewer as being "very delicately detailed in color" and his "paintings very industriously executed."[8]

Prestele was discharged from the army on September 27, 1823.[9] Two weeks later he was married (in Munich?) to Karolina, the daughter of Andreas Russ and his wife Elisabeth, née Schrey, the bride's father being a dealer in linen and garments. Very little about Karolina appears in the records but she was described, years later, as a gentle and patient wife and mother.[10] Of the couple's nine children, four died when young.[11]

A Religious Conversion and What Followed

JOSEPH PRESTELE HAD BEEN REARED A ROMAN CATHOLIC, BUT IN 1816 HE was attracted to the evangelistic sermons of Johannes Gossner, a learned and genuinely dedicated minister who had formerly been a Catholic priest. Gossner, as Prestele later reported, "preached about the gospel of Christ with a special force and warmth and thus that all our hearts were thoroughly stirred and each woke to the holiest decision to fashion our future life after the teachings of the gospel" and the example of Christ.[12] Gossner's teachings survived in what became known as the Gossner Group.

This was a time when old ways were being questioned and change was everywhere. The French Revolution, followed by the devastation of the Napoleonic Wars, had swept away ancient regimes and boundaries, and changed economies as well as ideas. In the process, old faiths were also destroyed.

As has happened so often during times of social chaos, many people sought escape from the trials of their real world in the comfort offered by the Bible and the teachings of Christ. This movement took many forms—including the search for direct and personal contact with the Holy Spirit—but all were a repudiation of the formalism and dogmas of the established churches. Those churches—and the governments with whom they were allied—sought to strengthen control over restless and fractious populations by repressive measures.

Prestele did not actually renounce his Catholic faith until much later, but when Gossner departed from Munich, leaving his little band of followers without a leader, Prestele took over the meetings. On occasion, services were held in his apartment at midday, and all who wished to come were welcomed. Although there was no one to give impassioned sermons like those of Gossner's, the gatherings followed the pattern of fellowship and spiritual communion that he had established with prayers, hymns, and religious discussions. There is no evidence that Prestele ever preached; instead, he was a quiet force, seeking conversions and providing spiritual guidance.

These innocent gatherings had no political overtones, nor were they in any way subversive, either to the established Roman Catholic Church in Bavaria, or to the Bavarian government. Nevertheless, such unorthodox actions aroused the suspicion of the authorities. They reacted swiftly—Prestele was closely watched by the police, and a dossier on his activities was assembled for the Catholic bishop in Augsburg.

In 1822, six years after he was first attracted to Gossner and his teachings, Prestele was arrested and sentenced to ten days in jail for having organized "a secret assembly under the pretext of worship in a house," a charge which Prestele stoutly denied, but to no avail. The police record covering his imprisonment includes a medical opinion that he was very weak and needed exercise in the fresh air. He was then permitted to spend a half-hour daily in the court of the police garrison. Apparently he was suffering from tuberculosis.

Other difficulties followed. For some years Prestele had received an annual stipend from the king, a customary way of encouraging the arts. In 1825, after the death of King Maximilian I, that stipend was terminated. Three years later, a budget reduction at the Royal Botanical Garden ended his appointment there, leaving him entirely dependent on income from his own work. Happily, his continuing friendship with the garden's director, Carl F. P. von Martius, and other botanists including Professor G. W. Bischoff of Heidelberg, resulted in a succession of commissions. Prestele was becoming favorably known among the painters and lithographers of Munich, many of whom were creating beautiful illustrations for the many important botanical books then being published, and whose work aided in making their city an early center for the new art of lithography.

When von Martius had returned in 1820 from his plant explorations in Brazil, he set Prestele to work drawing the specimens he had collected. It was a difficult assignment. Many of the dried plants were in miserable condition and, according to Prestele's friend, the theologian Magnus Joachim, von Martius had no talent in drawing so that his sketches were of little aid. Prestele often had to depend upon the author's verbal descriptions and his own imagination. Years later Prestele commented that he had made almost all of the outline drawings, in pencil without shading, used in von Martius's published work *Nova genera et species plantarum quas in itinere per Brasiliam Annis MDCCCXVII–MDCCCXX* (New families and species of plants found in a journey through Brazil in the years 1817–20). Few of the plates have Prestele's signature. However we know from the evidence of letters and references in various publications that much of his work in many books was unsigned.

Prestele also contributed to the illustrations of a work on Russian plants (*Icones plantarum . . . floram rossicam*, published in four volumes, 1829–34) by the German botanist and author, Carl Friedrich von Ledebour (1785–1851); and for the pioneering work on Japanese plants, *Flora Japonica* (published 1835–44), which resulted from studies made in Japan during 1823–30 by the Bavarian naturalist and traveler Philipp Franz von Siebold (1796–1866). (See Plates 4–7.)

In October 1834, Prestele submitted a petition to the government—specifically to the Royal Court of the Isar County Chamber of the Interior—seeking permission to buy a lithographic press so he could print his own work. He explained that "because of special personal reasons" he had moved to a Munich suburb which was some distance from the shop where his work had formerly been printed. "I am really put in a difficult situation because it is not possible, without great effort, danger, and cost, to transport the delicately worked stones

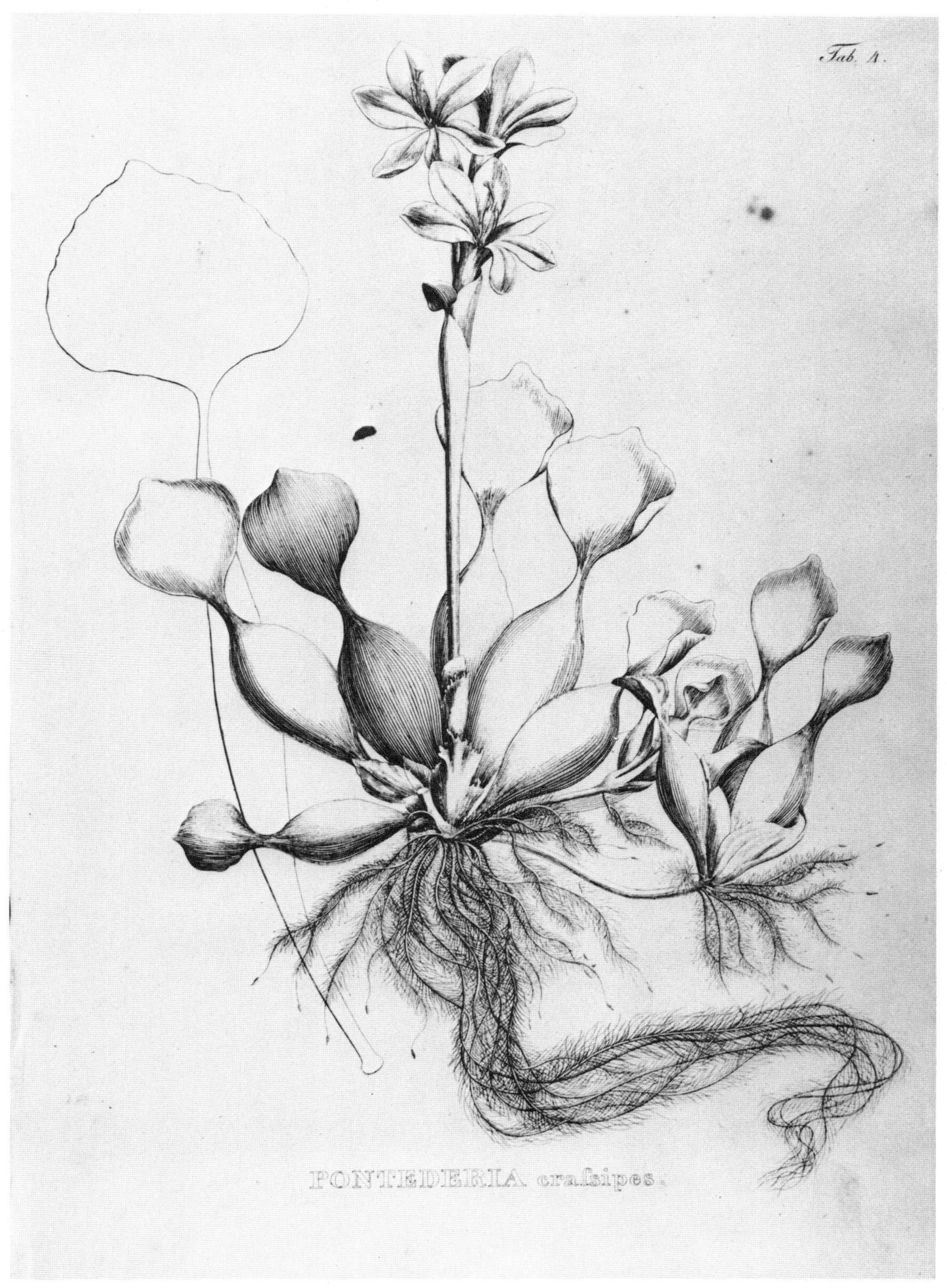

FIG. 7 "*Pontederia Crassipes.*" "Bischoff omnes del. Prestele omnes in lap. sculps." Lithograph; engraved on stone, colored. Plate 4 in Karl F. P. von Martius, *Nova genera*, vol. I, 1826. Courtesy, New York Botanical Garden.

Die wichtigsten

Giftpflanzen Deutschlands

in

lebensgroßen Abbildungen,

zur Warnung und Belehrung über die Gefahr

nach der Natur gemalt und auf Stein gravirt

von

Joseph Prestele,
Blumenmaler und Lithographen,

ausgewählt und beschrieben

von

Carl Soldan,
zweitem Lehrer am evangelischen Schullehrer-Seminar zu Friedberg.

Friedberg
in der Wetterau.

Druck und Verlag von Carl Bindernagel.

1843.

FIG. 8 Title page of *Die wichtigsten Giftpflanzen Deutschlands*, 1843, 43 × 27 cm. Courtesy, Gray Herbarium of Harvard University.

back and forth for corrections, etc." He asked to be allowed to buy a small press and set it up in his apartment. "Let me assure you in advance that I shall never misuse in any way this gracious license."[13] The urgency of his plea is apparent. The artist was then thirty-eight. He had to support his wife and five children, of which the eldest was ten.

His petition was denied.

Prestele and his struggles were observed with compassion by his friend Magnus Joachim, who lived with the Prestele family in Munich during 1828–29. Years later Joachim wrote that Prestele was a friendly, pleasant man, often sickly but always composed, never angry, and never despondent, who lived in very poor conditions. But "Often there was . . . quite miraculous relief from his dire distress. He worked all day to earn bread for his family"; he was a model "Christian father."[14]

About 1835, as Prestele was approaching forty years of age, he had another change in his religious thinking. Again he seems to have been influenced by a charismatic preacher, this one being the carpenter Christian Metz, a leader in the Community of True Inspiration [Inspirationists].

This sect had developed in Germany more than one hundred years earlier as one of the many protests against the formality and dogmatism of the Lutheran Church. The members—mostly artisans and peasants—believed that God spoke to His believers through individuals endowed with the miraculous gift of inspiration who were referred to as *Werkzeuge* (instruments).[15]

Prestele and his family joined the sect in 1837 and eventually he became an elder. This conversion represented a radical change in his religious thinking. Despite his leanings toward the evangelistic theology of Gossner and others, he had always remained a Catholic. Now he entered a mystical sect that condemned baptism and communion, believed in direct divine inspiration, and stressed an ascetic religious and celibate life.

As the Inspirationists grew in numbers, and in order to practice their faith more freely, the Community leased estates in Hessen-Darmstadt, including an old stone-walled building at Engelthal, formerly a convent, with its adjoining land. Here the Prestele family moved and here the artist produced some of his most beautiful botanical illustrations, the twenty-four plates for *Die wichtigsten Giftpflanzen Deutschlands* (The most important poisonous plants of Germany). The artist made the original drawings, engraved them on stone, and finally tinted the printed plates. The short text was written by Carl Soldan, a local schoolmaster who had become a friend of the artist's. The work was printed and published in 1843 by another friend, Carl Christian Bindernagel, who had a small printing shop in Friedberg, near Engelthal. (See Preface, note 1; and Plates 8–11.)

The following year Bindernagel also printed and published the artist's eight-

FIG. 9 "*Gemeine Waldrebe* [*Traveller's Joy*]. *Clematis Vitalba.*" Unsigned lithograph; drawn, engraved on stone, and colored by Joseph Prestele; 43 × 27 cm. Plate 17 from *Die wichtigsten Giftpflanzen Deutschlands* (Friedberg, Hessen: Carl Bindernagel, 1843). Courtesy, Gray Herbarium of Harvard University.

FIG. 10 "*Herbst-Zeitlose* [*Common autumn crocus, Meadow saffron*]. *Colchicum Autumnale*." Unsigned lithograph; drawn, engraved on stone, and colored by Joseph Prestele; 43 × 27 cm. Plate 20 from *Die wichtigsten Giftpflanzen Deutschlands* (Friedberg, Hessen: Carl Bindernagel, 1843). Courtesy, Gray Herbarium of Harvard University.

page, illustrated treatise, *Kurze praktische Anleitung zum Blumenmalen mit Aquarell-Farben* (Brief practical introduction to flower painting with watercolors). Prestele wrote and illustrated the work, which he signed as "Blumenmaler [flower painter] in Engelthal." (A copy of the pamphlet, containing two pages of "Four painted pattern plates," uncolored, is preserved at the Museum of Amana History. Unfortunately, the illustrations, drawn in outline and lithographed very lightly to permit more effective coloring, have faded to illegibility.)

The Community of Inspirationists continued to grow resulting in increased harassment by officials in Germany. It became clear that only through emigration—a possibility long discussed—could members hope to continue the free exercise of their religious beliefs. The decision to emigrate was spurred in 1842 both by a crop failure and, soon after, by the testimony of their leader, Christian Metz, that he had received a heavenly message ordering the Community to go west. The message included the reassurance "I am with you and shall lead you over the sea."[16] America, then, was to be their destination. In time, more than 800 members of the Community would find a new home there.

FIG. 11 The Wanderings of the Inspirationists in Europe and America. Lithograph; drawn, engraved on stone, and printed by Joseph Prestele, ca. 1850; 39.5 × 41.5 cm. Courtesy, Mrs. Lina Unglenk, Amana, Iowa.

Joseph Prestele made this plate after his arrival in America in 1843 to record the various places where the Inspirationists had lived in Europe, the ship that brought them across the Atlantic, and their new home at Ebenezer, in western New York. An earlier version of the print shows only the European scenes and the sailing vessel. (A copy of this first version is owned by the Amana Society, Amana, Iowa.) Later Prestele added the Ebenezer scenes for the version shown here.

2 To America, 1843

Ebenezer, New York, 1843–1858

IN AUGUST 1842, LESS THAN A MONTH AFTER MAKING THE DECISION TO EMigrate to America, a special committee of four men left Germany to choose a site for their new home. They landed at New York on October 26, where they learned that the United States government was moving the Seneca Indians from their Buffalo Creek Reservation in western New York, near the burgeoning city of Buffalo, west to the Indian Territory (which later became a part of Oklahoma). Their old reservation was to be sold.

After conferring with the agents for the sale of the property, the committee of Inspirationists traveled to the reservation, reaching there on a gloomy day late in November. A later historian, Frank J. Lankes, described their visit.[1] Part of the tract had been cleared by the Indians who still occupied their log houses, mission church, and council house. The land was gently rolling, the soil rich with humus. Most of the area was covered by the primeval forest, dim and solemn, with trees of immense size; great moss-covered glacial boulders were embedded in the earth. The scene, brooding and melancholy, was made even more somber by a dismal rain, yet the men were profoundly moved by what they saw. "From a sudden and rapid beating of their hearts [they] sensed an omen favorable to their desires—perhaps they had already found that place in the wilderness which they believed had been promised them." Their decision was quickly made. They agreed to purchase 5,000 acres; later adding 3,000 more. They called their new home Ebenezer, that is, "Hitherto the Lord hath helped us."

But even as plans for settling their purchase began, problems developed. The agents for the sale of the property were not able to get a clear title because of bumbling on the part of federal agencies, and thus could not convey the land to the Germans. As matters worsened, the state of New York and a group of Quakers, both anxious to protect the interests of the Senecas, became involved. The Indians added drama. At times they announced that they would not move from their ancient home; once in frustration and fury they summarily ordered the German colonists away.[2]

During this bickering, on May 1, 1843, only five months after Metz and his committee had first seen the property and purchased it (or so they thought), the first settlers from Germany arrived. The move had been carefully planned: other parties came at regular intervals and found that accommodations were ready for them when they reached Ebenezer.

Few of the colonists could speak English but they learned enough about the situation from their leaders—and from sensing the tension about them—to become anxious. Still, with seeming indifference to the confusion and confrontations with the Indians still living there, the settlers proceeded to cut timber, build houses, lay out streets and roads, and in other ways undertook the enormous labor of creating a settlement of villages and outlying farms in the "Indian Wilderness."[3] Their leaders sought to meet each crisis as best they could; councils were held with the threatening Indians; new contracts and additional payments were made; they consulted with their lawyers, with the agents involved in the sale, and officials representing the state and national governments. In the midst of all this, the Presteles arrived, part of a group of sixty-seven that reached Ebenezer in October 1843, some fifty-six days after leaving Engelthal. Their crossing had been a rough one. Joseph later wrote to his parents back in Germany that he had been violently sick during the entire voyage.[4] His berth below decks was in a space crowded with other immigrants and near the door to the hold from which came an almost overpowering stench of rotting vegetables and other perishables. At times, weakened by constant nausea, he despaired of living to see America. Finally, the ship's captain, a kindly man, moved Joseph to his quarters where he was provided good drinking water and plenty of chicken soup. When the ship docked at New York on September 30, Joseph was much improved.

From New York the immigrants went by boat to Albany, thence by a slow train to Buffalo reaching there on October 7. They were met by one of the brethren from Ebenezer with a wagon to take them to their new home, some two hours away. It had been raining hard all day and the roads—miserable at best—must have been almost impassable. Later Joseph merely commented, "The roads were not good."

Toward evening the Prestele family neared their destination. From a low rise they saw the village of Middle Ebenezer, then in the process of construction, and Joseph learned that a house had been built for him. It was a most happy piece of news, for he had expected they would have to live in a dark log cabin for a year or so. When they reached their new house on School Street they found it a frame

FIG. 12 House occupied by the Joseph Prestele family, 1843–58, in Middle Ebenezer (now Gardenville), New York, as it appeared about 1865. Artist unknown. Lithograph; drawn on stone. Illustration in *The Illustrated Historical Atlas of Erie County, New York.* (F. W. Beers & Co., 1880), p. 98–a. Courtesy, The Buffalo and Erie County Historical Society, Buffalo, New York.

The clapboarded frame house is shown as it appeared about seven years after the Presteles had moved to Iowa and when it was occupied by a miller whose gristmill had been built beside it. Although considerably changed the house still survives on School Street in Gardenville.

building of comfortable dimensions, freshly painted white and made cheerful by many windows. It was on a high foundation because there was danger of flooding from Big Buffalo Creek that ran in front of it, along the other side of the street. Although the yard was in disorder and work remained to be done inside, the exhausted family thought the house "friendly and nice."

The more than 800 Inspirationists who eventually came from Germany to Ebenezer were a diversified group but represented the composition of the society, the majority being artisans and farmers.[5] Some were well-to-do, many were poor. A number, including Joseph Prestele, had to borrow from the Society's treasury

FIG. 13 "*Mittel Ebenezer von der Westseite.*" (Middle Ebenezer from the west.) Unsigned pen-and-ink drawing, colored, by Joseph Prestele, Sr., ca. 1850; 34 × 19.5 cm. Courtesy, Mrs. Lina Unglenk, Amana, Iowa.

The large building (center) was the Middle Ebenezer Meeting House, its entrance facing west. The logs in the millrace (left) and in the foreground were on their way to the sawmill. The building partially shown in the extreme left was a mill. The buildings are described and located by Frank J. Lankes in his *Ebenezer Community of True Inspiration* (Gardenville, N. Y.: published by the author, 1949; reprinted 1982?).

to finance the trip. Soon after his arrival at Ebenezer, the artist described himself as a "poor man with a large family and am indebted to the Society for my passage across the ocean."[6]

At Ebenezer each family received a house with living and sleeping quarters and a large garden. The houses had no kitchens as all meals were taken in communal dining halls. Men and women wore somber, practical clothing as directed by their leaders; any vain display in dress was forbidden. "Theirs was only a church brotherhood that fed, clothed, and sheltered them physically as well as spiritually," Frank Lankes summarized.[7]

During the next few years, the Presteles took part in the miracle that transformed the wilderness into an orderly pattern of villages and farms and factories. By the end of 1846, less than four years after the first settlers had come, the Inspirationists had cleared some 1,000 acres, built three villages—two with thirty-five houses each, and a third with eight houses, a sawmill, and a grist mill.[8] Eventually, they built four hamlets in all: Middle Ebenezer (now Gardenville), Upper Ebenezer (Blossom), Lower Ebenezer (the present Ebenezer), and New Ebenezer, which was located on Big Buffalo Creek between Middle Ebenezer and Upper Ebenezer. Most of these villages were within the present township of West Seneca, Erie County, New York.[9]

These villages grew in irregular patterns according to needs and circumstances. The village of Middle Ebenezer was located on Big Buffalo (or Seneca) Creek in order to utilize its water power for a grist mill. A cemetery was designated and orchards were planted at the outskirts of the village. New Ebenezer developed in a similar way; it was also on Big Buffalo Creek and had the Community's water-powered woolen mill.

The houses built by the colonists varied in size and, except for a few brick structures, were made of wood with heavy timber frames and gable roofs. In design and construction they were somewhat simplified versions of traditional Germanic houses but they showed some Americanisms, such as clapboards on the outside walls, a concession to the extremes of the American climate. Further insulation was provided by the German method of filling the frames of the outside walls with brick rubble, and by placing rolls of straw and mud in the spaces between the joists supporting the first floors. The interior woodwork was left in its natural state and after frequent scrubbings developed a satiny sheen. The settlers did most of the construction themselves, aided by some outside labor.[10]

Joseph Prestele was fascinated with his new world, for he had many interests and his artist-trained eye missed nothing. He was intrigued by American inventiveness and love of machinery. He admired American tools for their innovative designs and excellent materials. American felling axes were very different from German ones and they both delighted and terrified him. "I am frightened every time I hold one in my hand . . . Several of our men have hurt themselves using them." He marveled that American woodsmen cut down forest trees in a fraction of the time it took the immigrants with their German axes.

Nails, Prestele wrote his parents, were cheap in America because they were machine made, as were window frames which came in all sizes, ready to be inserted in the walls. He learned that frame houses could be built in about six days and, in a comment suggesting prefabrication, reported that a completed house "can be moved with machinery to the desired spot." His own house, Prestele said, had been made of large timbers cut at the sawmill and put together on the site. "The outside is finished with smoothly planed narrow and thin clapboards painted with white oil paint." The inside partitions were lathed and plastered.

In the letters that Joseph wrote to his relatives in Germany within the year after his arrival at Ebenezer, he said little about his family. Nowhere is his wife mentioned, either during the voyage—which must have been harrowing to her too—or later when her work in organizing their new home and taking care of their children's needs was never-ending. Of the children he had very little to say and less that was affectionate. In September 1844 he wrote that he was "well satisfied with my Gottlieb; he is very diligent and nearly as tall as I am." His youngest son Henry, born at Engelthal and then about six years old, was a "skilled and clever youth but lively and active." Elise enjoyed her work in the communal kitchen and was "tall and very thin." Karolina [Caroline] Elise, born in 1835, was very small and weakly but she was well and "her work is satisfactory." His eldest son Joseph, Jr., was a disappointment. He was unruly, his father wrote with some bitterness. He could have added that the young man was restless and irresponsible, for soon after the family had reached Ebenezer, he left his home and the Community of Inspirationists, going first to Buffalo and then, rather mysteriously, to New York City.

IN 1846, THREE YEARS AFTER THE COLONISTS HAD BEGUN TO DEVELOP THE tract, the last of the Indians departed for the West and the Inspirationists came into full possession of their purchase. They had won out in the end, largely "through the enduring quality of their faith."[11]

During that same year, the membership accepted the "Constitution of the Community of True Inspiration at Ebenezer,"[12] a document which sought to establish practical as well as ethical regulations for the conduct of the Community and its members under the new and difficult conditions created by life in America. The members were finding that the freedoms of America were beginning to endanger the solidarity of their Community. In Germany, persecution had forced individual members to seek the protection of their religious Community. In America, where no such pressures existed and where people were free to live and work wherever they chose, the lure of the outside world was affecting the members of the community and the temptation to leave was becoming evident. There was also the growing problem of supporting the less fortunate members of the Community—of maintaining what was, in effect, a classless society among its members.

Their constitution, among other provisions, established a limited form of communism and strengthened church control over the lives and actions of the members.[13] ("All of our outwardly organized arrangements are closely related to our religious rules and principles so that I cannot say much about one without touching the other," Prestele once wrote his father.)[14] The leaders maintained that this move conformed to a message from the Lord—as revealed by Metz—that it was "His most holy will that everything should be and remain in common."[15] Land, buildings, machinery, livestock, and other such assets were declared common property. No one received wages or profit; all work in the communal enterprises was done without payment in money. Instead, workers received credits which could be used in the Community stores. Surplus products were sold and the income went into the Community fund. The Society was responsible for food, shelter, medical care, and finally, the burial of its members. Individuals, however, owned their household furnishings, clothing, and small tools.[16]

As the task of building diminished, the everyday rhythm of life at Ebenezer became more routine. Frank Lankes, who lived in Ebenezer many years after the Community of True Inspiration had moved to Iowa, imagined the sounds and movements and odors of the villages and their outlying farms as they most probably existed.[17] Wagons would clatter occasionally across plank-floored bridges; there were intermittent whines, rumbles, rasps, and other mechanical noises from the few mills but they did little to break the tranquility. The odor of smoke hung like incense in the air. As twilight came the light of candles and open wick oil lamps gleamed from houses; darkness and silence soon followed. On the Sabbath, the workaday noises were replaced by the sounds of nature—of insects, and birds, and water tumbling over the mill dams, and the wind in old trees. This drowsy quiet was broken only by the *Glockenhaus* ("belltower") bells calling the people to worship, and by the passage of an occasional horse-drawn rig.

As time passed there were some defections from the Community at Ebenezer, particularly among the young men, but in general the system worked well and the Community survived until 1932.

Joseph Prestele's individualistic profession as an artist and his contacts with learned men was an anomaly in that tightly controlled church society, with its emphasis on the practical. In America, the leaders did not encourage his talent and work, thinking it too "worldly" for a religious community, although, as we shall see, they were interested in the income his work brought into the Community's treasury.[18] But if Prestele ever felt that he was something of an outsider, he gave no hint of it in his few letters that have survived. And certainly, his faith never seems to have wavered. "God himself rules and regulates all the outer and inner religious incidents in our community," he wrote to his father.[19]

Joseph Prestele, like so many German immigrants, abhorred disorder and confusion. Because of that trait, and because of his love of natural beauty and his training as a gardener, he began almost immediately after his arrival to "prepare my house and its surroundings real nice, most of it before or after the communal work hours to avoid criticism that I had used community time in doing unnecessary beautification."[20] Around his house he made a narrow, raised bed, edged with thick boards. Here, in the spring of 1844, he planted several kinds of morning glories, climbing beans, and sweet peas, which were trained to grow up lattices made of thin boards and nailed to the walls. By the end of the summer the vines had almost reached the rooftop. Elsewhere he planted two peach and three pear trees. On either side of his house he made a garden planted with flowers and also some lettuce whose seed a friend in Engelthal had given him and which grew well in the rich soil. He looked forward to the time when his yard could be fenced so that he could have a garden in front of his house too.[21]

Prestele's first work assignment for the Community was to take charge of the apple orchard left by the Indians and to develop a nursery for propagating more fruit trees. But before he could begin the nursery, he had to clear the land. He described that strenuous and often dangerous task in May 1844 to Carl Christian Bindernagel, his friend in Friedberg, Hessen, who had published his book on poisonous plants the year before.

> *Sometimes the trees are so overgrown with vines, thick as an arm, also other plants, that they form a regular roof and often I do not know where to start my work. Since we are still short of people and helping hands for the many, many chores, I mostly have to work alone . . . and dear God gives me strength and health. Of course in the evenings I am very tired, especially my arms, for the ax, the hatchet, and the spade are no brush handles . . . I thank God for His mercy in saving me from injuring or even losing my eyes, nose, hands, and feet.*[22]

In his rare moments of leisure, Prestele found delight in the unfamiliar wonders to be seen in the American forests and meadows, the wild flowers, the birds with their bright plumage and cheerful songs, and the silent, graceful snakes which he described in letters to his parents.

An Art Career Resumed

AS THE MONTHS PASSED, AND THE NOVELTY OF LIFE IN AMERICA BEGAN TO fade, Prestele began wondering if gardening, working in the nursery, and clearing land of trees and brush was the best use that could be made of his time. He had given up sketching and painting fruit and flower pictures when he came to Ebenezer and had been satisfied with his outdoor assignment. But during the summer of 1844 he began to realize that he was not physically able to do such heavy work.[23] We suspect too that he missed his art work and the pleasure and excitement that had come from his association with men like von Martius, Soldan, and others. He began thinking that at Ebenezer there were many beautiful wild flowers which he could paint and perhaps sell for the benefit of the Community. But he held those thoughts to himself in the belief that if God wished him to paint again, He would make a sign.

A week after he first had these thoughts, something happened that convinced Prestele that God had, indeed, given the sign. As he recounted in a letter to his father on August 25, 1844, two of the "first brothers" (elders?) came to him and said that they had decided he should take up his art again. A few weeks later he was busily sketching, aided by his son Gottlieb. It took a while for Joseph's stiff and worn fingers to become accustomed again to using the brush and pencil, but he made the start. At first he and his son collected and sketched the most beautiful and interesting wildflowers they could find, pressing examples for reference in making watercolors during the winter months. Since he couldn't identify most of the plants and didn't know how he could get such help in America, he wrote his friend Joseph Gerhard Zuccarini, professor of botany at the University in Munich to seek his aid, with the intention of sending him dried plant specimens and seeds later.

Then, in January 1845, he acted upon another idea. Before Prestele had left Munich, Professor Zuccarini had given him a letter of introduction to Asa Gray, professor of botany at Harvard University. Prestele, however, lost the letter during his travels. Now, with the help of a friend who could write in English, Prestele prepared a letter to Gray. It may have been the most important letter the artist ever composed. Addressing it to "Asa Gray, Esq., Professor of Botany, at the University, Cambridge near Boston, Mass.,"[24] Prestele spoke of his work at the Royal Botanical Garden in Munich as "a drawer and painter of plants." He mentioned some of the work he had done for various botanical works, including the three volumes on Brazilian plants by von Martius, whose studies would have been well known to Gray. Prestele went on to say that during the previous summer of 1844 he had collected some of the wildflowers and herbs growing around Ebenezer with the thought of laying a foundation for "a botanical work or Flora americana, by painting them according to nature." He was troubled, however, because he was unable to identify the plants and asked Gray if he could assist. He closed with: "I beg to tender you my services whenever you will afford me an opportunity to render any."[25]

The letter from Prestele was of great interest to Gray for he had begun work on publications that required the services of a skilled botanical artist, particularly one like Prestele who could engrave on stone. During the exchange of letters that followed, Prestele sent examples of his work which Gray enthusiastically described to the scientist John Torrey as a "most superb set of drawings, both of cultivated and some native plants, exceedingly well done. Also specimens of his work in cutting on stone, which he does admirably."[26] Torrey, an older friend and mentor, was a botanist, chemist, and physician. He was then teaching at Princeton University and, concurrently, at the New York College of Physicians and Surgeons.

Gray's comment referred to Prestele's skill in the art of engraving on stone, a rarely used technique similar to engraving on copper that had the advantages of "facility of production; accuracy of drawing; minuteness of details; and clearness of impression."[27]

The process was one of a number of lithographic techniques developed by the inventor of lithography, Aloys Senefelder, by which an image can be transferred to a lithographic stone (a very fine-grained type of limestone) for printing. All lithographic methods are based on the antipathy of oil and water or, in this case, the antipathy of greasy printer's ink and water. The use of the word "engraving" in lithography is something of a misnomer, for the lines forming the image are not cut into the stone, but through a thin coating of gum arabic to the stone's surface, using engraver's tools. This temporary coating serves to protect the surface while it is being worked and helps make corrections easy. Once the image has been cut through the coating, the entire surface is covered with a greasy or oily substance (such as linseed oil) which fills the lines and penetrates the stone. The gum is then removed by washing, but the design—which will accept the greasy printing ink—remains. After further treatment and inking, the stone is ready for printing.[28]

DR. TORREY NEEDED THE ASSISTANCE OF AN ARTIST-LITHOGRAPHER AND wrote Prestele asking about his availability and his charges. The artist replied in German, explaining to him, as he had to Gray, that he knew no English, and there was then no one at Ebenezer who could write a letter in that language. He said that his charges depended entirely on the amount of time required. Outline drawings "with the necessary analyses of the flower & plant could be had for $2." The engraving on stone cost from $3 to $5.[29] To better explain his charges and show the quality of his work, Prestele sent examples, including six pages of his black-and-white line engravings from the *Botanical Terminology* of Professor G. W. Bischoff, of Heidelberg, on which he noted his charges. These had ranged from $10 to $27 a plate.[30]

Torrey, impressed by what he had learned, sent the artist drawings to engrave. Prestele had the "needles" for such work but not the lithographic stones. Pending the arrival of an overdue shipment of stones from Germany, he ordered a sufficient number for his immediate needs from the G. & W. Endicott lithographing firm in New York City.[31]

The Endicotts proved to be helpful friends for many years. The two brothers, George and William, were highly skilled lithographers and they had made their shop the center for that trade in New York. Their customers found them

FIG. 14 Bouquet Tied with Ribbon. Unsigned lithograph; engraved on stone by Joseph Prestele, Sr.; 30.7 × 25.7 cm.; no caption. Courtesy, Mrs. Adolph Schmieder, Middle Amana, Iowa.

This fine example of Joseph Prestele's skill as an artist and engraver on stone was apparently made to be colored and sold for framing as a decorative print, not as a nurserymen's plate. Neither this print nor the colored example (Plate 84) has a caption, and the subject is not included in any of the three Prestele catalogues (lists) of their nurserymen's plates reproduced in Appendix C.

agreeable and progressive. Prestele was able to buy from them the supplies he needed from time to time. They were patient with his occasional complaints and seemingly anxious to make amends for the mistakes or oversights of their employees.

Meanwhile, Gray had also been in further correspondence with Prestele. Gray outlined the publications he had under way and asked if the artist could assist with the illustrations. He too inquired about the artist's charges. Prestele replied (April 5, 1845) that he would "feel gratified to be allowed to take a part" in Gray's projects. He would stop gardening and other such manual labor "as soon as I am engaged in your work . . . I shall not charge you more than I have asked from Mr. Torrey, that is Seven Dollars for four Tablets [plates], which is as low as ever any man can ask." Prestele went on to say that it was difficult to quote prices in advance because the work varied so greatly, but that those for whom he had worked in the past were "always satisfied with the prices I asked after the work was completed and I am sure you too will be satisfied." He mentioned that the charge for "the lithographies of the German poisonous plants [examples of which he had sent Gray] . . . and for which only a few analyses were necessary, I received 2 to 3 Dollars. For the plates of Bischoff's *Terminology* I received 5 to 40 Dollars* for the most beautiful."

Further correspondence with Gray followed. On June 12, 1845, Prestele wrote that he could use the facilities of the Buffalo firm of lithographers, Hall & Mooney, until he received stones and bought a press. He suggested that as soon as he finished one piece of work for Gray he would have a proof made and send it to him for approval and correction. "Everything will go as it will go," he added fatalistically. He was nervous about taking up engraving again for the delicate work required absolute muscular control. It "will be somewhat difficult at the beginning; for two years I have not worked at this—with needles on stone—but I did hard labor where my hands became stiff and heavy."

Gray put the artist to work on his current project, a publication that was to illustrate interesting North American plants, chiefly those growing in the Harvard Botanical Garden, which was under Gray's care. Drawings of the selected specimens were to be made by Isaac Sprague,[32] a native New Englander, some fifteen years younger than Prestele. With Gray's encouragement Sprague had moved to Cambridge where he worked under Gray's close supervision.

Sprague's career, then just beginning, was to be long and distinguished. In time he would become one of America's greatest botanical artists. Like Prestele

---

* *An error in the letter? This is a much higher figure than Prestele charged for other difficult work.*

he was infinitely careful in his work, refusing to be hurried by the impatient Gray and sometimes, one suspects, being irritated by his employer. Sprague could be moody and given to complaining of physical ills, which won little sympathy from Gray whose vigorous health made it difficult for him to be patient with the frailties of others. But the two men understood and respected each other and worked together effectively. Both were cordial to Prestele, accepting him as an experienced and skilled artist.

Although Prestele was handicapped in communicating with Gray, both by the language barrier and by the distance of Ebenezer from Cambridge, the artist's frequent reports and queries—and Gray's terse replies—kept things moving, although not always at the pace that Gray wished. He always seemed to have too much to do, and too little time.

As Prestele came to know Gray better and found encouragement and aid from his friend in Cambridge, the artist's letters began to include reports on his daily work and problems, his search for supplies needed in his engraving and printing, and, at times, references to his charges. In financial matters Prestele was low-keyed, somewhat subservient, appreciative of favors shown him, and infinitely persistent. In a letter of September 12, 1845, he spoke of charging $5 for engraving two plates and offered to color them at "5 per hundred in a handsome style." Later he wrote Gray he would have to charge $4 each for engraving plates of the *Gaillardia*, *Obolaria*, *Oakesia*, and *Sullivantia* because of the considerable work involved. "I hope, therefore, that you will not find the charge . . . unreasonable; the remaining four plates I charged the same price as for the Thermopsis [$2.50] although some of them are a little more difficult than the latter."

Through these first commissions from Torrey and Gray, the pattern of Prestele's work in America was almost accidentally established. In Europe he had had opportunities to work in a variety of media, as we have seen—watercolor, oil, and lithography. But the immediate need of Torrey and Gray was for engravings made from the drawings of Isaac Sprague, whose work they knew and trusted, and for tinting of the printed plates.

Because so much of Prestele's work, both then and during the years that followed, was as an engraver, it might be assumed that he was essentially a copyist who contributed little or nothing beyond his technical and manual skill to the final design of the plates. This was not true. Where scientific objectives would not be compromised, Prestele could and did alter the sketches of others to make the plates more artistically pleasing and more informative. There is a clarity of line that makes his art distinctive and his sensitivity as a colorist is apparent in all his work.

On January (8?), 1846, Prestele wrote Gray that he and his seventeen-year-old son Gottlieb were busy coloring fifty of the plates. Some were difficult, like

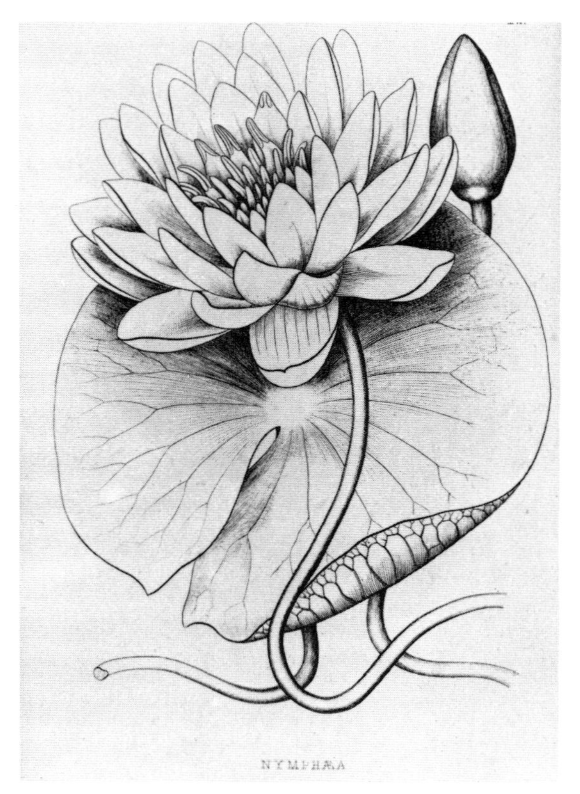

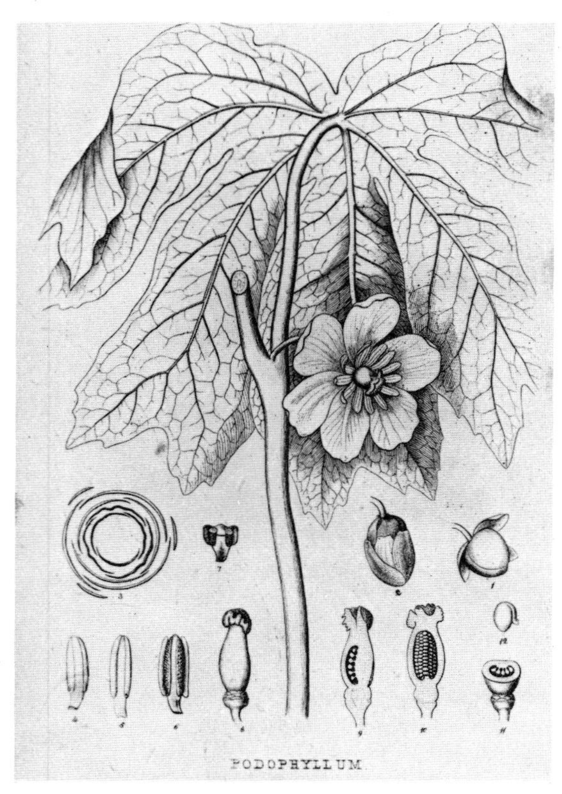

Left, FIG. 15 "*Nymphaea*." (*Nymphaea odorata*. American Waterlily.) *Right*, FIG. 16 "*Podophyllum*." (*Podophyllum peltatum*. Common Mayapple. Mandrake.) Both, lithograph; engraved on stone; 23.5 × 15 cm. Plates 42 and 35, respectively, of Asa Gray's *Genera florae Americae*, vol. I. (New York: George P. Putnam, 1848.) Courtesy, the Botany Library, Smithsonian Institution. Inscriptions are given only on Plate I of each volume. For volume I: "J. Prestele omnes sc. Sprague omnes del. Printed by G. & W. Endicott, 59 Beekman St. N.Y."

the plate for the "*Gaillardia amblyodon*, [which] gives us much trouble, and because it is such a handsome plant we wish to do it well, which is the reason of being under the necessity to work slow, and thus we are not able to paint more than five or six plants [i.e., plates] per day, and therefore would ask you if $10 per hundred would be too much?—The *Oakesia conradii* is likewise a very difficult plant and is certainly worth at least $7 per hundred tablets [plates]."

In 1846 the first installment of Gray's treatise on North American plants was published in the *Memoirs* of the American Academy of Arts and Sciences. The title: "Chloris Boreali-Americana. [Flowers of North America.] Illustrations of New, Rare or Otherwise Interesting North American Plants, Selected Chiefly from Those Recently Brought Into Cultivation at the Botanic Garden of Harvard University." It included fifty-six pages of text and ten plates, nine of them colored. The design of these illustrations, and the subtle coloring, ranging from the rich

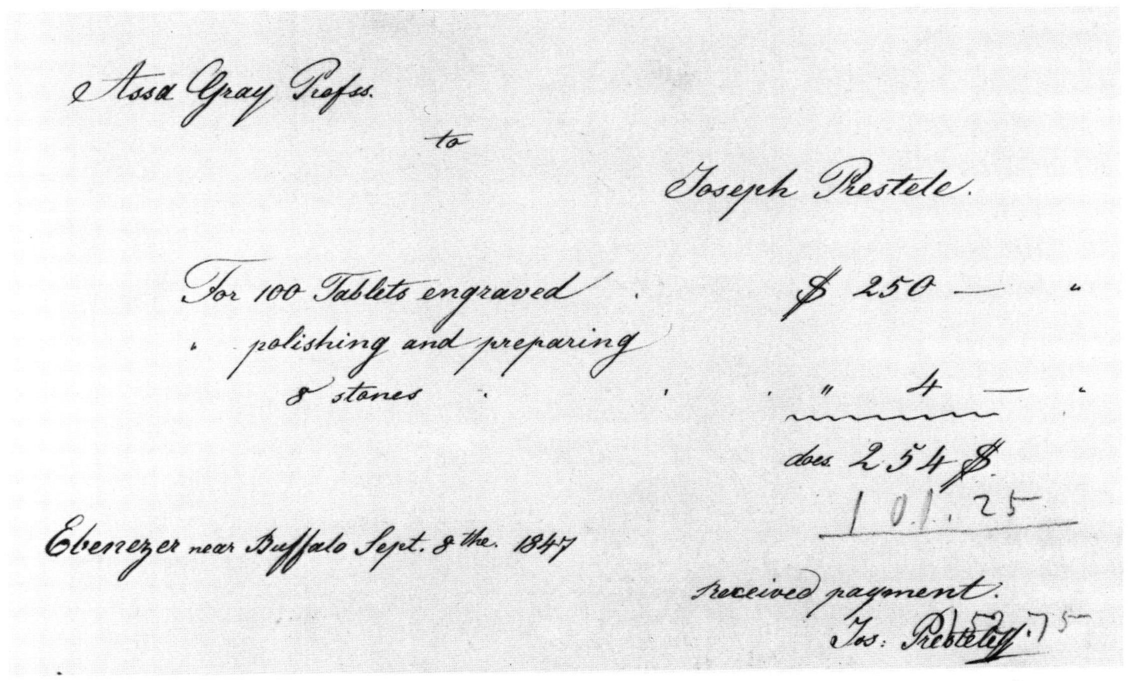

Assa Gray Profss.
to
Joseph Prestele.

For 100 Tablets engraved . $ 250 — "
" polishing and preparing
8 stones . . . " 4 — "
does 254 $.
101.25

Ebenezer near Buffalo Sept. 8the. 1847

received payment.
Jos: Prestele 152.75

FIG. 17 Joseph Prestele's bill to Asa Gray, September 8, 1847, for engraving on stone the illustrations in volume 1 of Gray's *Genera florae Americae*, 1848. Courtesy, Gray Herbarium of Harvard University.

texture of the *Gaillardia amblyodon* (maroon gaillardia) to the delicate shading of the *Oakesia conradii* (*Uvularia*; merrybells), resulted in plates of exceptional beauty.

Although Gray had planned originally to continue the "Chloris," by then he had developed an even more grandiose idea, the *Genera florae Americae boreali-orientalis illustrata* (The genera of the plants of the United States illustrated by figures and analyses from nature), in which a single species of each genus of plants in the United States would be selected for description with an authoritative illustration. Further work on the "Chloris" was dropped.[33] Sprague and Prestele were pressed into action and the new work moved along rapidly. By September 8, 1847, Prestele had engraved 100 plates for the book and sent Gray a bill for them at $2.50 each, and $4 for "polishing and preparing 8 stones." The first volume of the *Genera florae Americae* was published in 1848 with 100 uncolored illustrations, drawn and engraved in outline. Volume 2 followed in 1849. Then that project, like its predecessor, was abandoned. Gray had envisaged ten volumes but could not raise the funds to continue it beyond two.[34] The project was probably costing more and taking longer than he had anticipated, but then Gray was always impatient. At times he grumbled that Sprague was slow, and although he admired the quality of Sprague's work, he resented the stubborn determination of the Yankee to work at his own pace. Sprague charged $6 a drawing, but, as Gray wrote a friend, he took so much "time and care" with them that he did not earn day wages.

Production Details

PRESTELE HAD MANY DIFFICULTIES WORKING AS AN ARTIST AND LITHOGRApher in rural and remote Ebenezer. He needed supplies of various sorts; lithographic stones, special paper, inks, and coloring matter, for all of which he had specific requirements, based upon his experience in Europe and the high standards which he was determined to maintain. He often experienced difficulty in obtaining the items he sought. Frequently his work had to be printed by distant firms, which involved the risk of shipping fragile stones with even more fragile impressions drawn on them. In addition, there were the high costs involved, a matter to which he gave the closest scrutiny.

From Prestele's letters to Asa Gray and others, we learn a great deal about the practical aspects of his operations. His search for the right kind of lithographic stones was particularly vexing. In 1845, after coming to Ebenezer, Prestele ordered stones from Germany, explaining to Gray (June 12, 1845) that the stones could be bought there for a fraction of what they cost in Buffalo. "One stone of 2 feet in length and 1½ [feet] in width would cost me in Frankfurt 1 Dollar 20 cents, here in Buffalo 6 to 8 Dollars . . . I shall find out in New York . . . prices of various sizes."

The stones ordered from Germany had finally arrived by August 1845, but more were soon needed as demand for Prestele's work increased. One assignment came from the Boston schoolmaster and botanist George Barrell Emerson (1797–1881), then engaged in completing his *Report on the Trees and Shrubs Growing Naturally in the Forests of Massachusetts*, which had been commissioned by the state legislature. Emerson apparently asked Prestele to engrave on stone seventeen plates of drawings by Sprague.[35] For this work Prestele ordered four more stones from G. & W. Endicott, New York, as he reported to Gray on October 5, 1845, adding that "after having received them I can fix the price [of the stones] for Mr. Emerson." (Included in the same order were four stones for Gray's work.) Emerson's *Report* appeared late in 1846. It included the seventeen uncolored plates, beautifully drawn and engraved. Four were published without any credits at all, the balance with only the name of the printer, G. & W. Endicott, New York City.

The eight stones ordered from Endicott, weighing roughly 500 pounds, were sent by railroad instead of by canal boat, which would have been cheaper. Prestele objected to the charge. Eventually he won a $3.50 reduction on the $20.00

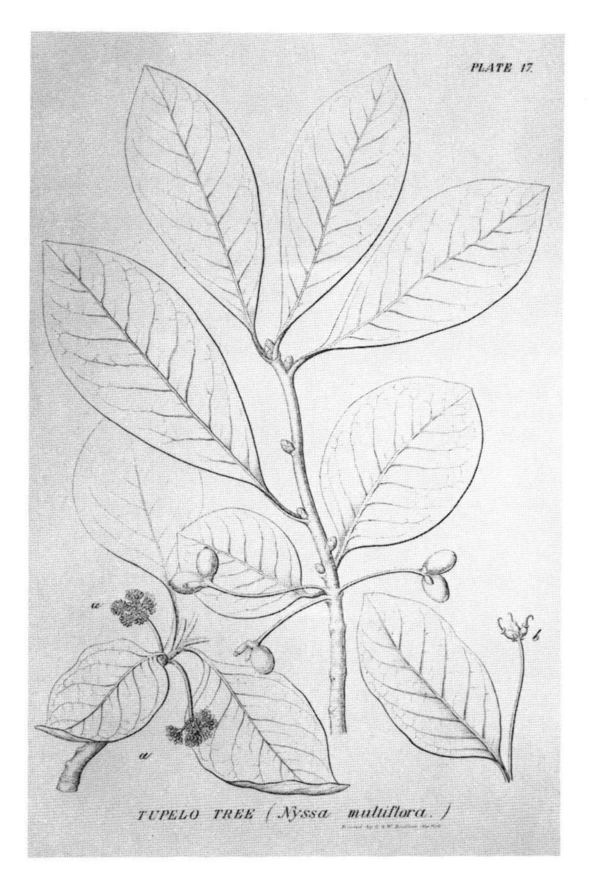

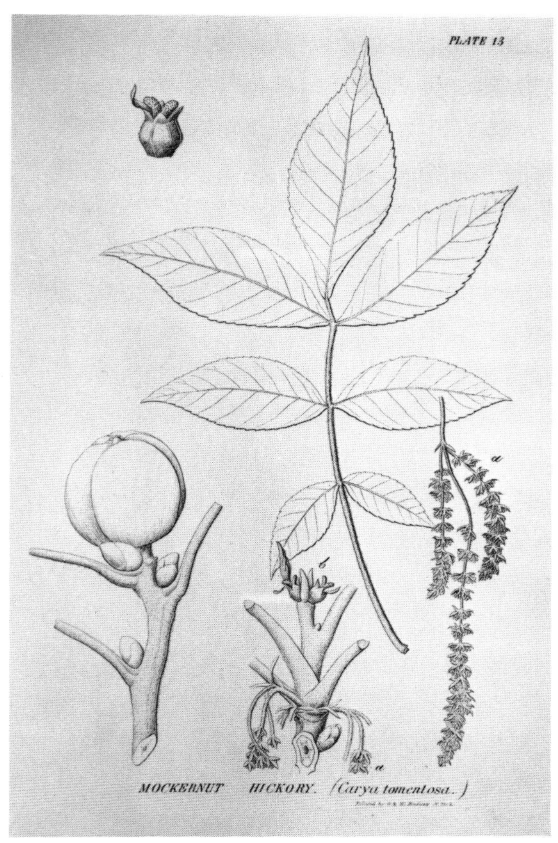

Left, FIG. 18 "*Tupelo Tree (Nyssa multiflora).*" (*Nyssa sylvatica*. Black Tupelo, Sour Gum, Pepperidge.) *Right*, FIG. 19 "*Mockernut Hickory (Carya tomentosa).*" (*Carya tomentosa*. Mockernut Hickory, Bullnut, White Hickory.) Both plates inscribed: "Printed by G. & W. Endicott, N. York." Lithograph; engraved on stone by Joseph Prestele from a drawing by Isaac Sprague; 23.7 × 15.5 cm. Plates 17 and 13, respectively, of George B. Emerson, *A Report on the Trees and Shrubs Growing Naturally in the Forests of Massachusetts*. Published Agreeably to an Order of the Legislature, . . . (Boston: Dutton and Wentworth, State Printers, 1846).

express charge, and a $10.00 reduction on his bill from Endicott, who conceded that his "carman" had erred in sending the order by rail. These revised costs Prestele passed on to Emerson and Gray. (His clients paid for, and owned, the stones used for their work, and apparently they also paid for the paper.)

Prestele also lamented that Endicott had only sent him "white stones, which are badly calculated for such work," being softer in composition than dark gray stones.

Lithographic stones were a major expense and Prestele was thrifty with them. Some seven stones that he, and later his son Gottlieb, used have been preserved and they show how skillfully the Presteles made use of their surfaces. The size of these stones, which were used for nurserymen's plates, varies from 11 by

17 inches to 20 by 16 inches. Generally, the Presteles drew images on both sides of a stone and, depending on the design, crowded from two to four images on each surface. Since individual designs were printed at different times, each with wide margins, the printing required considerable skill.[36]

As his work increased Prestele frequently had difficulty obtaining paper of the size and quality required to produce the best impressions and to take color well. In the autumn of 1845 Gray sent him two samples, one of which the artist said could be used for impressions to be colored, and the second for uncolored prints. For colored plates Prestele preferred "sized" paper,[37] which didn't absorb as much ink as unsized paper and hence produced the light impressions that were desirable if the plates were then to be tinted.

Some of Prestele's clients were insistent that only the best paper be used for their work. William Starling Sullivant (1803–1873), the Columbus, Ohio, bryologist, urged Gray (who was assisting in making Sullivant's arrangements with Prestele) to be "very particular about the paper, pure white very smooth & even texture, & rather thick—Nothing sets off an engraving like good paper—60 or 70 copies will answer me."[38] The author's insistence on fine paper and fine engraving was justified. When his "Contributions to the Bryology and Hepaticology of North America: Part I" (communicated to the American Academy of Arts and Sciences on August 12, 1846, and later published in 1848) appeared, Sir William J. Hooker, director of Kew Gardens, London, commented, "The plates are peculiarly excellent, drawn by Mr. Sullivant himself . . . and beautifully engraved."[39]

BEFORE COMMERCIALLY SIZED PAPERS FOR LITHOGRAPHIC WORK WERE REGularly available, Prestele—like other lithographers—had to size his own paper to improve its quality for printing and coloring. The process called for coating the paper with a glaze, or glue, using gelatin, starch, or gum, either singly or in combination.[40] What formula Prestele used is not explained in his letters, and his general references are confusing. He spoke of the work as "glueing and pressing the paper." It was his nemesis. "It is very troublesome and dangerous work and takes considerable time," he lamented to Sprague on February 11, 1851. Some sheets wrinkled when drying. He could not do the work for less than 25 cents a hundred sheets. Prestele was delighted when he found that sized paper of good quality was available in New York City for $9 a ream. (Some years later the price had increased to $12 a ream.)

By the summer of 1846 Prestele had decided to do his own printing. Toward that end he bought a lithographic press in New York and wrote Gray that he was awaiting the arrival of a friend from Germany who was a printer on stone and who would bring a supply of lithographic stones with him. Prestele did commence printing—presumably his German friend arrived—but Prestele's few references to printing for several years suggest that he had difficulties in setting up the process. Early in 1850 (January 5) he wrote Gray that he did not think his printer could "strike off the stones which are drawn with crayon to your full satisfaction" and recommended shipping them to Gray so they could be printed in Boston. He added, however, that he would like to print the stones he engraved "with the pin." But by February 16, 1850, his situation had changed, for he wrote Gray that he was "now capable to perform all kinds of printing work as good as anyone, therefore I wish to engage my printer for I had great trouble and heavy expense in order to get him here and also for the press and I would be glad if I could get the printing of all those impressions which become colored here but the uncolored or plain ones might be printed there in Boston." Prestele continued to provide printing, but he was never able to get all his clients to let him do theirs. Much of the printing for Gray, Torrey, and other clients of Prestele's continued to be done elsewhere.

Before Prestele had facilities for printing, he shipped the prepared stones to the firms doing that work. How the stones were crated to prevent marring the images on them while in transit is not explained in his letters. During 1845 and into 1846, the firm of Hall & Mooney, conveniently located in nearby Buffalo, did much of Gray's printing, but to Prestele's increasing dissatisfaction.[41] Despite their best efforts, as he wrote Gray in February 1846, they lacked experience in such work, "and their printer does not take pains enough in his laying on the paint [ink]; which is the cause of having so many Impressions come off so pale. I have been myself in Buffalo and tried to assist them, but since I only understand that kind of work in theory, and not in practice, I could not alter things to my views. I hope however for the better and think after a little more exercise in such work all will turn out nice and good." But that hope was not realized. Thereafter, much of Gray's printing was done by the Endicott firm in New York, which Torrey was using.

Prestele seems to have been pleased with the quality of the Endicott's work, but not—at times—with their apparent carelessness in shipping and billing. In October 1849, he believed the firm had wrongly billed him $21.50 for corrections and lettering on a job they were doing for Gray. Prestele had engraved the stones, and as he wrote Gray on October 17 [1849], "I struck off the proofs of most of the stones . . . and did likewise the corrections and the many alterations myself, of which some were very difficult. I think the proofs should have been done by Mr. Endicott himself, because he has the printing of the work."

The Endicott firm was asked to letter some of the engraved stones received from Prestele, for he was always unhappy about performing such work. "The text or letters on the drawings I would rather leave Mr. Endicott to have engraved," he wrote Torrey in August 1845, "as I think I cannot do it handsome enough." To Gray he once confided, "I am not so well practised in lettering."[42] Sometimes, under pressure, he reluctantly took on such work, as on February 23, 1858, when he wrote Torrey he would letter his plates in the future, "for I wish not that you get trouble with it for my sake. I shall try to do it as good as possible it is for me, you but will please write well the names on the drawings." Perhaps the artist's aversion to lettering is one of the reasons why he signed very few of his plates.

PRESTELE'S CLIENTS FOUND IT INCONVENIENT TO HAVE HIM WORKING SO FAR from them. It created delays and made communication difficult. Gray tried to get the artist to move to Cambridge to be near him, as Sprague had done some years before, but he was no more successful in his efforts than William Starling Sullivant in Columbus, Ohio, had been. That wealthy businessman, who was fast becoming famous as an amateur botanist and bryologist, wanted Prestele there. He explained to Gray that if Prestele were in Columbus, it would "save me a great deal of labour in making the drawings & Prestele could do the engravings a great deal better if I could be at his elbow often."[43]

Prestele also refused Torrey's invitation. "I cannot move to New York," he wrote Torrey on April 7, 1845. "I have a family with 5 children and live among a separate Ctn. [Christian] Community."

In 1849, when Gray again felt it imperative that he, Sprague, and Prestele be able to work more closely together, he invited the artist to visit him in Cambridge. Prestele declined, using his unfamiliarity with English as the excuse. Gray then suggested that Sprague visit Prestele at Ebenezer, to which Prestele reluctantly consented on August 10. "I fear he would not stay many days with me on account of not being able to accommodate him with such lodging as would do for him since we are without those accommodations as yet, which a Gentleman like Him ought to enjoy, and to which he is used to. But if he was to run the risk concerning these things, I will be ready for his visit, and wait on him as good as our circumstances here will allow."

Sprague made the long trip to Ebenezer in the autumn of 1849. Despite the language barrier the two artists found much of common interest, and a friendship developed that continued undisturbed throughout the years of their working relationship. On October 17, shortly after Sprague's visit, Prestele wrote Gray that he hoped Sprague had returned in good health. "I was very glad to become acquainted with him personally, and I wish I had been able to converse with him in his own language." Prestele and Gray never met.

The rapport between Sprague and Prestele was strengthened by the visit and resulted in even more effective cooperation. Sprague made the drawings and gave Prestele a description of the leading colors, or "laid them on" himself. If Sprague or Gray found the colors on the proof weren't accurate, Prestele altered them. Prestele engraved the drawings on stone and drew proofs which were checked by Gray, who returned them to the artist for correcting or printing. If Prestele was not to do the printing, either because of his client's wish or for technical reasons, he sent the stones to the designated firm. Sometimes the printer had to make corrections on the stones or return them to Prestele for reworking. After printing, the plates were returned to Prestele for coloring, along with the stones. It was a clumsy system, consuming much time and increasing the costs, but somehow it worked.

Nothing makes clearer the cordial relationship between Prestele and his clients, Gray and Torrey, than their harmonious financial dealings. Only once did Gray try to get Prestele to lower his charge, and he did not persist against the artist's patient stand. Sometimes Gray felt the artist charged too little. Late in 1846 he wrote Prestele, apparently offering to increase his payments for engraving, but on December 4 Prestele replied declining the offer. "I feel myself very much indebted to you for favors received already and since I live very much retired and frugal my expenses are generally met without much difficulty, I am glad when you keep me busy and my son."[44]

Torrey, like Gray, was also considerate of Prestele in financial as well as in other matters. In September 1851, after Torrey had arranged for Prestele to be given some government work, he told the artist that his fee of $5 for engraving the plates was too little, and that $15 should be charged.[45] Later [January 14, 1858], when Washington was slow to pay, Prestele appealed to Torrey: "Can't you tell me how I may get the money, as you know I rely always on your kind advice in all respects." On another occasion, when engraving a plate for Torrey, he wrote [July 5, 1855] that he would be satisfied with whatever Torrey paid him: "I know you always do what is right."

"A High Standard of Excellence"

BY 1847, ONLY TWO YEARS AFTER HE HAD FIRST WRITTEN TORREY AND GRAY asking for work, Prestele had established a considerable reputation among American botanists and horticulturists. It was a remarkable achievement for a recent immigrant who was just beginning to speak and write English, and who lived in a religious community far from the intellectual centers along the eastern seaboard. We have no record of all the commissions he received, but no doubt one that came during 1847 must have given him special pleasure, not only because it meant that his abilities were recognized by the nation's leading horticultural society, but also because of the background of the invitation.

The commission came from the Massachusetts Horticultural Society in Boston, asking Prestele to engrave and color four plates for volume 1, number 2, of its *Transactions*, [46] a new publication with which the officers of the society were determined to set a high standard of excellence. The plates in the first number, however, had been an embarrassing fiasco. The society had engaged William Sharp,[47] a Boston artist and lithographer, to illustrate the first number in "a very superior manner." It was thought that Sharp could use the new process of chromolithography, which he had introduced to Boston, to make colored illustrations of prize fruits and flowers more quickly and at a lower cost than was possible by the older, largely hand methods used by artists like Prestele.

When the long-delayed first number of the *Transactions* appeared on July 18, 1847, it included an apology by the society's Committee of Publication. They had expected "plates in a style of excellence much superior to that of those which now accompany it." After much trouble and disappointment the committee decided "that the process of chromolithing, in its present state, is not adapted for a work of the character which it is determined to stamp on the *Transactions* . . . or to give even a faint idea of the beautiful drawings made by their artist, Mr. W. Sharp." They added: "Future numbers shall appear in a very different style."

The source of their disappointment was the coloring of the plates. Andrew Jackson Downing, the influential editor of the *Horticulturist, A Journal of Rural Art and Rural Taste*, commented in his magazine that while the plates were richly colored and more carefully executed than those in "Mr. Hovey's serial," the *Fruits of America*,[48] they lacked "that fidelity to nature, and delicacy of tint, which characterize the best English and French coloured plates, done by hand."[49] Robert Thompson, of the London Horticultural Society, was both more emphatic in his criticism and more explicit. In a review in Downing's magazine, Thompson proclaimed the coloring "*decidedly bad*." Among other shortcomings, he noted that the wrong shade of yellow had been used on both a pear and its leaves, and an apple was colored an unnatural red. He ended his diatribe with "Judging from the specimen before us, chromolithing will not answer for fruits. False coloring tends to mislead—a plain representation is far preferable."[50]

"Chromolithing" [chromolithography], which Thompson had denounced, was then in its first stages of development. The process was one by

FIG. 20 "*Tyson Pear.*" "Lithd. of J. Prestele." Lithograph; engraved on stone, colored; 26 × 17.5 cm. In *Transactions of the Massachusetts Horticultural Society*, vol. 1, no. 2 (1848). Courtesy, New York Botanical Garden

which a picture could be printed in color, instead of having to be printed first in black-and-white (or some other single color) and then tinted by hand. "Chromos," as they came to be popularly known, were produced on a series of lithographic stones, or zinc plates, the number depending upon how many colors were used in a particular job. Each stone had a different portion of a picture drawn on it and given one color. The series of stones were so arranged that the different colored impressions blended into a complete colored picture. This method of printing was largely mechanized and capable of rapidly turning out many copies. It would have numerous advantages when perfected. In time it made possible the widespread use of colored illustrations, posters, and other commercial work and was capable of producing very fine art work.

The disappointment of the Massachusetts Horticultural Society with Mr. Sharp's chromolithographs led to a hasty review of plans for the next issue. The Committee of Publication renounced any further adventurism with chromolithography and turned to the tried and true hand techniques used by Joseph Prestele. They were not disappointed.

The second number of the *Transactions* came out early in 1848 with four plates which Prestele engraved on stone and colored, and for which he may also have made the original drawing, and printed. They were all that the society had hoped for. They illustrated fruits then popular, the "Dix Pear," "Andrews Pear," "Downers Late Cherry,"[51] and "Tyson" pear—the last signed "Lith^d. of J. Prestele." Downing applauded their quality. "These plates are finely executed, being engraved and coloured by hand, and are far superior to those in the first number, which were done in the *chromolith* process, which the Society has abandoned."[52]

Pleased with the beauty of Prestele's colored engravings and the favorable comment they elicited, the society's publication committee engaged the artist to prepare four more plates for the succeeding third number of the *Transactions*, dated January 1852. These illustrations, without signatures, were the "Beurre D'Aremberg" pear, "Dearborns Seedling Pear," "Red Astrachan Apple," and "Heathcot Pear."

With that number, publication of the *Transactions* ceased. Albert Emerson Benson, in his history of the society, said that the need for such a publication no longer existed. (See note 46.)

Plates for the Book That Was Never Published

IN THE SPRING OF 1848 PROFESSOR JOSEPH HENRY, THE FIRST SECRETARY OF the newly created Smithsonian Institution, wrote Asa Gray asking his friend to undertake a study of the economic and other uses of forest trees in our country, which the Smithsonian would publish.[53] He intended it as a major work in the institution's new program for the "increase and diffusion of knowledge among men," as required by Joseph Smithson's will endowing the Smithsonian. More specifically, Henry planned the work as part of the Smithsonian's new series of books—Reports on the Progress of Knowledge—which, although intended for a general audience, were not to be light entertainment. Instead, he asserted, these publications would require "attention and thought to understand them."

Henry realized that much of the success of the tree book would depend upon its illustrations. He knew the capabilities of Isaac Sprague and Joseph Prestele through their work with Gray, and he knew that their contributions would be essential to the success of the venture.

Gray had little enthusiasm for Henry's project at first, but wanting to help his friend and the infant Smithsonian, he agreed to undertake the book. As he thought further about the project, it began to excite him. "We have gradually enlarged our ideas . . . much beyond the original plan, as to the figures," he enthusiastically reported to John Torrey,[54] his older friend and fellow botanist.

Henry first announced the tree project in 1848.[55] He believed that a variety of public uses would be served by a work that combined botanical data with such practical information as the comparative strengths of wood and the methods of the lumber industry. The study was to be limited to native trees within the "United States proper" or, as he sometimes put it—and as it was later officially titled—*The Forest Trees of North America*.

In his *Annual Report* for 1849, Henry enthusiastically described the project in more detail. It was, he asserted, "The most important report now in progress." The book would appear in two octavo volumes of text, with an atlas of quarto plates—the first part to be published in the spring of 1850. The introductory section was to include the present state of knowledge, "divested as much as possible of all unnecessary technical terms." He reiterated that as the work "will be adapted to the general comprehension, it will be of interest to the popular as well as the scientific readers."[56]

Soon after Gray had reached his understanding with Henry he started Sprague and Prestele working on the plates for the projected book. In this assignment they followed their usual division of work: Sprague made the drawings of plants from dried or living specimens, and Prestele reproduced them, either by "engraving" them on stone or, in some instances, by drawing them in "chalk" (a greasy crayon) on stone, the more familiar lithographic process.[57]

Gray, meanwhile, had begun writing the introductory section, thus hoping to get the project well underway before departing on a year's study trip to Europe. But as usual, Gray was overly optimistic. Too many things had to be done before he left, as he wrote his English friend George Bentham.

> *I have to finish this* [Ferdinand J.] *Lindheimer collection, finish* [Augustus] *Fendler's, distribute and study* [Charles] *Wright's collection when I get it, carry the "Botanical Text-Book" through the press, rewriting and expanding it (thus far I have made it all over), write the first volume of an elaborate report on the Trees of the United States for the Smithsonian Institution, in fact a Sylva, with colored plates by Sprague (which*

I could not resist taking in hand, as that institution promised to bring it out, and handsomely, at their expense), and give my course of lectures in the college from March to June. When all of this is done I can cross the Atlantic.[58]

On April 15, 1850, a few months before Gray left for Europe, he and Henry signed "Articles of a Convention," an agreement more personal than legalistic that would haunt them for the rest of their lives.[59]

This curious document included a review of Gray's understanding with Henry about the "Report on the Trees," the work on it accomplished to date, and Gray's responsibility for the correctness of the bills that Sprague might submit to the Smithsonian while Gray was in Europe, their total not to exceed $300.

Also included in the agreement was a progress report on the illustrations, but that summary is so confused that one is left uncertain as to how many different subjects were actually involved.[60] In addition, it was agreed that during Gray's absence Sprague was to finish only about eight more drawings, yet he was also to continue making drawings at "the proper moment of all the trees intended for the second volume, that fall in his way." It was stated that ten woodcuts had been engraved from Sprague's drawings under the microscope, illustrating the growth and formation of wood (these, apparently, were never published).

If the "Articles of a Convention" were intended to clarify the confused operation, and to quell concern about the slow progress of the book and its mounting costs, it failed. During the following months while Gray was in Europe, it became increasingly clear that matters were not going well. In preparation for a meeting of the regents of the Smithsonian on January 1, 1851, the troubled secretary sought a detailed report from Isaac Sprague on the production of the illustrations, the payments made to date, and the anticipated expenses for the months before Gray's return. His request ended with the rather plaintive—and "any other information in your power."[61]

Plans continued to be changed. Instead of three volumes, as originally announced, it was decided to issue the work in two volumes, each with 100 plates. Most surprising for a book intended for general distribution, the size of the edition was eventually limited to about 400 copies—or so it would appear—for only about that number of each plate seems to have been produced.

As the months passed there were unexpected delays. The production of the colored plates, from the first sketches to the finished state, took much time and patience on the part of the artists who also had other commitments. Perhaps in answer to a query about his slow production, Prestele explained to Dr. Spencer Fullerton Baird, assistant secretary of the Smithsonian, that coloring a plate like that of the *Aesculus glabra* required much care. "I can do very few a day."[62]

Prestele charged $4 for engraving each of the tree plates. His prices for coloring them varied considerably according to the length of time required. Such work was infinitely tedious, required the utmost attention, and was a strain to eyes, fingers, and back. Some details were almost microscopic; brush strokes had to be precise. The selection and use of colors for such work was highly technical. In his little work *Brief Practical Introduction to Flower Painting with Watercolors* (see p. 35), published in Friedberg in 1844, Prestele gave some practical rules. These included painting a leaf "with all the shades of green, the use of carmine when painting roses, the blue for bell-flower, the yellow for narcissus, wallflower and saffron, and finally, the most difficult of all to portray, the white flower."[63]

Between 1849 and 1854, Prestele's lowest charge for tinting plates for the tree book was 6 cents although, as he wrote Gray on January 5, in his new and uncertain English, "my earn the day is very small, and since I have made my prize [price] for engraving very low already." Gray, seeing the costs mount, sought a lower figure, but Prestele would not go below six cents, quoting the adage, "Like pay, like work." Prestele added that if Gray still could not afford to pay more than "6 Cts per piece then we must do them according." For some plates Prestele asked considerably more, as with the magnolias. These were to be the first illustrations in the tree book and Gray took particular pains with them, insisting again and again on alterations. The patient artist quoted a price of 15 cents each for coloring the final version, adding, "I am Dear Friend very sorry that I troubled you so much in this case."[64]

Prestele sometimes felt that Sprague's designs could be improved, either for artistic or scientific reasons, and proceeded to make the changes, generally minor ones. During Gray's absence in Europe, Prestele altered a detail in the design of the American Linden drawing. He explained to Sprague on January 22, 1851, that one of the leaves hung behind the blossoms and he thought that when the leaf was colored it would spoil the looks of the pale yellow blossoms in front of it. When Gray returned he scrawled on the proof, "Badly done. It shall be done over."[65] However, the plate as eventually issued, has a leaf partially behind a cluster of the blossoms. (See also Plate 36.)

While Gray was in Europe he left Sprague in charge of the tree project. As part of that responsibility, Sprague forwarded the accounts to the Smithsonian for payment. In April 1851, Sprague was told that Prestele's latest bill was being paid but that "it is not expedient for us at present to meet any farther drafts until Dr. Grays return from Europe."[66] Although this was contrary to the agreement the Smithsonian had made, Prestele said he was willing to wait patiently with his bills, but he hoped that the work could continue in the meantime. In November 1851, after Gray's return, when the Smithsonian was again short of funds and there was a threat of temporarily halting work on the tree book, Gray offered personally to send Prestele an advance so that production could continue. The appreciative artist said he could wait until "next January [1852] then [the] Professor will pay it at once I think." But by February [1852] Professor Henry had not paid and Prestele had become "very anxious for it." Eventually he was paid but he continued to have difficulties collecting his accounts. In December 1852 the artist wrote Gray that he was "anxious waiting for the bill." Gray forwarded the letter to the Smithsonian with a scribbled note, "Please pay the bill of Prestele. You have it."[67]

At various times Gray queried Prestele about the possibility of printing the plates for the tree book in colored ink, perhaps in the hope of reducing costs. When Gray asked if the impressions could be struck off in green ink, the artist explained that although this would work all right for the leaves, it would not for flowers and fruit if they were to be tinted another color. "But," he added patiently, "if you wish to have them done or printed in any hue, I am ready to colour as you wish them."[68] When Gray raised the question again in February 1850, Prestele reminded him that the Massachusetts Horticultural Society had "tried to avoid painting [i.e., hand coloring] by the process of chromolithography but all in vain, up to this day, for that art will never become a competent Substitute for coloring with the hand." In January 1852 Gray asked if there would be any advantage in printing certain plates in color; again Prestele denied it. He explained that the subtle coloring required was very difficult to apply even by hand.

Gray does not seem to have been convinced by Prestele's arguments, because colored ink was used in printing the outlines of a few plates for the tree book: these were *Magnolia grandiflora*, *M. glauca*, *M. umbrella*, and *M. auriculata* (Plates 1, 2, 3, and 4). They were drawn in crayon on stone, Plate 2 by Prestele and the three others by Antoine Sonrel, an artist and engraver in Boston who was known for his skill in producing scientific plates.[69] Tappan & Bradford printed them in green ink; red ink was also used for a section of Plate 3. The advantage of using the colored inks is not apparent; indeed, these plates lack the brilliance of those engraved and printed in black and it is possible the colored inks are to blame. Prestele colored at least two of these plates (3 and 4). He completed about 400 copies of each subject and since Gray was in Europe, forwarded them to Sprague in two shipments, December 1850 and February 1851. Plates for the *Magnolia acuminata* and *M. cordata* were also printed by Prestele about the time the project was being terminated, but those plates were never colored, and it is not known what became of them.

Gray's search for a less expensive method of producing colored plates was part of the effort being carried on throughout the Western world to develop a low-cost method of color printing. He was among those who sensed the potential of chromolithography as a cheaper, faster, and less troublesome way of reproducing illustrations than by the hand methods used by the Presteles and others. Indeed,

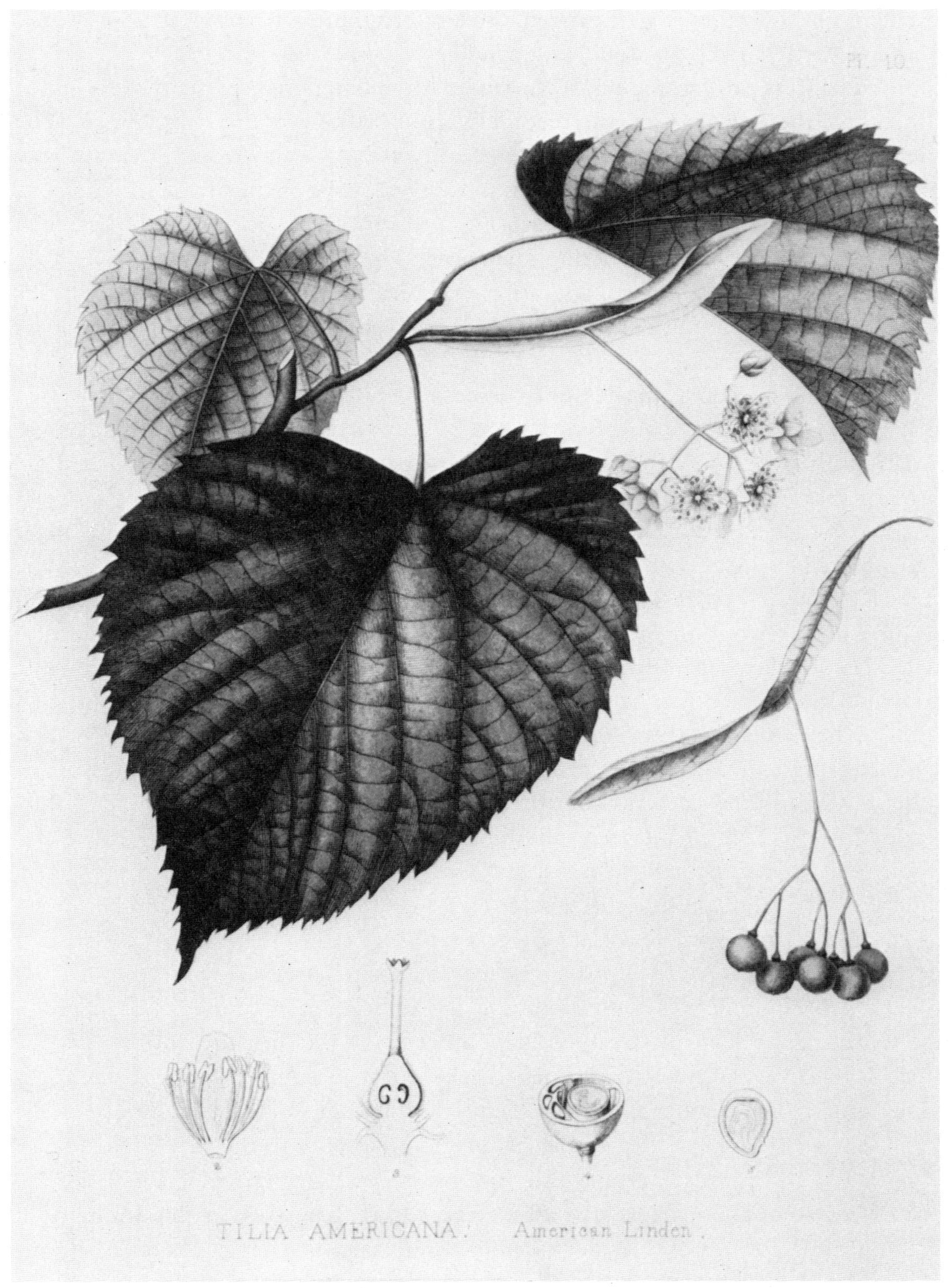

FIG. 21 "*Tilia Americana. American Linden.*" Unsigned lithograph; drawn by Isaac Sprague; engraved on stone, printed, and colored by Joseph Prestele; 34 × 27 cm. Plate 10 for the projected *Forest Trees of North America* by Asa Gray.

as chromolithography was improved, the old techniques of "engraving" on stone, drawing with "chalk" on stone, and hand coloring passed from use.[70]

Progress on the tree book continued to be hampered by many things, not the least of them being Gray's exceedingly critical editing of the artists' work. The men were painstaking but Gray was not easily satisfied, and would preemptorily order a plate to be redone. Sometimes a plate had to be changed again and again, often requiring a new engraving, printing new proofs, and coloring the accepted versions to satisfy Gray's exacting standards. The quality of the work resulting from such a conscientious collaboration was very high, but deadlines were never met, costs steadily mounted, and the funds of the Smithsonian proved inadequate for such an undertaking.

Professor Henry, in the Smithsonian's *Annual Report . . . for 1850*, explained that the work on trees had been delayed by Gray's absence abroad, but that the illustrations were going forward and "the first part will probably be published the present year." He closed with the rather ominous comment that the costs would be greater than first anticipated, so that the publication of the entire work "must necessarily be spread over a number of years."[71] Henry again made excuses in the *Annual Report . . . for 1851*, adding that Gray had returned from Europe "and will resume the preparation of the drawings, as soon as the funds of the Institution will admit of the expenditures." He reiterated that the work "will form a valuable contribution to the botany and economical and ornamental arts of our country."[72]

Thereafter progress was sporadic, and work finally came to a silent and embarrassing halt. Gray never produced the text; there was always the pressure of other duties and the attraction of other ideas and projects. He may have felt handicapped by his lack of information about many species of trees in the West and Southwest. It is possible that he lost interest. Like others of Gray's early projects, the tree book "trailed off in the mists even while prophesying more and better yet to come."[73] And Henry found the project far more costly than he had anticipated or the Smithsonian could afford; apparently he too was willing to drop it.

"I believe there is a mind among the Officers at Washington to abandon the work," Prestele wrote Gray on November 13, 1855, "because nothing is done since several years, and also you have spoken of it in some letters written a few years ago."[74] But neither Gray nor Henry was willing to admit failure publicly. Even in his 1856 *Annual Report*, his last printed reference to the tree book, Henry maintained an optimistic front, although his efforts were rather strained. He repeated many of the excuses he had offered before. He stressed the difficulty in making the drawings of trees for the illustrations because they could only be done at the various stages in the annual leafing, flowering, and fruiting of the trees. He said that all the drawings would be original and that many had been finished. He expected the first part of the work to be ready for the press during the year (1857),

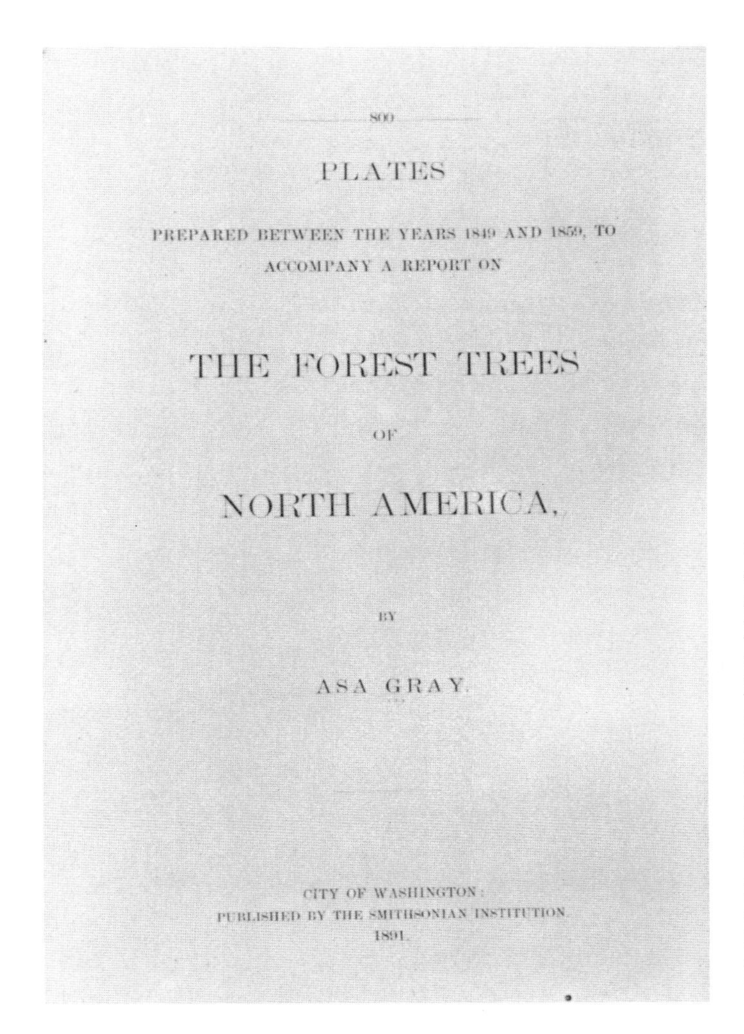
800

PLATES

PREPARED BETWEEN THE YEARS 1849 AND 1859, TO
ACCOMPANY A REPORT ON

THE FOREST TREES

OF

NORTH AMERICA,

BY

ASA GRAY.

CITY OF WASHINGTON:
PUBLISHED BY THE SMITHSONIAN INSTITUTION.
1891.

FIG. 22 Title page of the portfolio containing twenty-three plates distributed by Secretary Samuel P. Langley of the Smithsonian in 1891 to several hundred botanists and libraries. These plates are the only tangible result of a project which the Institution had launched almost fifty years earlier, to compile a major work on the forest trees of North America. It was to have been the Smithsonian's first major publication in its series of "Reports on the Progress of Knowledge," but for a variety of reasons, the project was never completed. Author's collection.

and that the work would be published in quarto form which, he said, was the best adapted for the illustrations.[75]

Most of Prestele's work on the tree book was done from 1849 through 1851, or so it would appear from the surviving letters. In December 1850 the artist shipped to Sprague 1,600 completed plates; in February 1851, 1,329 more, the two shipments representing ten subjects.[76] In September 1851 he wrote Sprague that he was sending the rest of the colored plates. Thereafter Prestele received only minor jobs connected with the project, and even they ended in 1854. Finally, on October 1, 1858, Prestele reported that he had sent six stones, with twenty-two engravings on them as Gray had directed, to the Boston lithographers Tappan & Bradford.[77]

The finished plates and the lithographic stones on which they had been printed, all the property of the Smithsonian Institution, remained at Gray's house until after his death in 1888, when they were sent to the Institution. In 1891—as we saw earlier—Dr. Samuel P. Langley, then secretary of the Smithsonian, distributed a few hundred sets of the twenty-three plates, together with his two-page history of the failed venture, to leading botanists and museums of the world. The gift was, as he put it, a memento of Gray who had given so much of his life and labor to the study of American trees. In the *Annual Report . . . to July, 1891*, Langley further stated, "The illustrations, so far as furnished, were skillfully drawn by [Isaac] Sprague, and were reproduced on stone by [Antoine] Sonrel, [Joseph] Prestele, and others; the impressions being carefully colored by hand."[78]

Thus in 1891 the episode of *The Forest Trees of North America* was finally closed. It had been a frustrating and embarrassing experience to both Henry and Gray. The latter's widow, Jane Loring Gray, in selecting and editing her husband's letters, which she published in 1894, included only scant references to the affair. Langley's comments in the Smithsonian's *Annual Report* and in the sets of plates which he distributed in 1891 attracted little attention. After the sets of plates disappeared into library collections, they—and the plan for the tree book—were apparently forgotten.

Reports on the Botany of the Far West

AFTER PRESTELE HAD CLEARED AWAY THE FEW LOOSE ENDS REMAINING FROM the tree project, he received only occasional work from Gray, not because of any differences between the two men but because Gray was busy with projects that didn't require Prestele's engravings. Nevertheless, Prestele was kept busy with various commissions which his growing reputation attracted. Some of these were of little consequence, such as the four watercolor sketches of fruit that he supplied for reproduction in volume III of *The Agriculture of New York*, by Ebenezer Emmons, M. D. (part V, 1851, of *The Natural History of New York*.)[79]

Other commissions of more importance came through the offices of Dr. John Torrey, that good friend who was always willing to lend the artist support when problems arose, and to advise and encourage him. Some of the work involved Torrey's own writings. One such was the illustration accompanying Torrey's "Observations on the Baris Maritima of Linnaeus," published in the Smithsonian's series, Contributions to Knowledge, April 1853. It was a short article on an obscure plant for which Sprague and Prestele produced a minutely detailed black-and-white engraving.[80]

Torrey was also responsible for Prestele's being commissioned to engrave botanical illustrations for some of the official reports of army explorations in the Far West. When Prestele undertook that assignment he was fully aware of its important scientific aspects.

By 1850—after the war with Mexico, the California gold rush, and the American settlement of Oregon—a proud and exultant young nation had expanded to the Pacific. A series of U.S. Army explorations and surveys were carried out to learn about the vast new region that had been acquired and, in addition, to search for a transcontinental railroad route and also to resolve problems relating to the new boundary with Mexico. In time the names of such officers as Captain Howard Stansbury, Major William H. Emory, Lieutenant Edwin G. Beckwith, and Lieutenant Amiel W. Whipple—among others—would be known for their journeys into the recesses of western canyons and mountains and for the information they brought back about that hitherto unknown region. Since their orders included assembling natural history specimens and data, the material they collected, together with their sketches and notes, provided the basis for carefully drafted scientific reports. Professor Torrey was put in charge of identifying much of the botanical material collected, assisted at times by Asa Gray, Dr. J. M. Bigelow, and Dr. George Engelmann, among other leading American botanists. The army contracted with Prestele to engrave many of the botanical plates.*

Through this government work, Prestele, in his self-effacing way, continued to be associated with important men and important activities. A wealth of botanical discoveries poured in from newly explored corners of the Far West, but if the dried plant specimens, collected in strange and unimaginably distant places, sometimes at the risk of death from Indians, excited his imagination, there is no hint of it in his prosaic letters. Yet the illustrations that he and his fellow artists produced were among the first to accurately portray the landscape, the natives, and the plant and animal life of a region which had seemed, to the civilized world, a mysterious unspoiled paradise, more fable than reality.

In September 1851, Prestele began engraving drawings of botanical specimens collected on Captain Howard Stansbury's expedition to the Great Salt Lake region in 1849–50. The work was difficult because the drawings were "thickly packed with figures," as Prestele wrote.[81] Then followed work for many other reports. Not all of Prestele's plates can be identified because so many illustrations appeared without signatures. However, some clues to his work can be found in his letters to Torrey, in Torrey's letters to Gray and others, and in Torrey's prefaces to botanical reports. In some instances Torrey specifically acknowledged the assistance of artists, as in Lieutenant Amiel W. Whipple's *Exploration and Surveys for a Railroad Route . . . in 1853–54* where, Torrey stated, "All the engraving has been done on stone by Prestele, who excels in this branch of the art."[82]

Prestele's letters referring to his work on the Whipple plates are particularly

---

* *A partial checklist of these reports is given in Appendix B.*

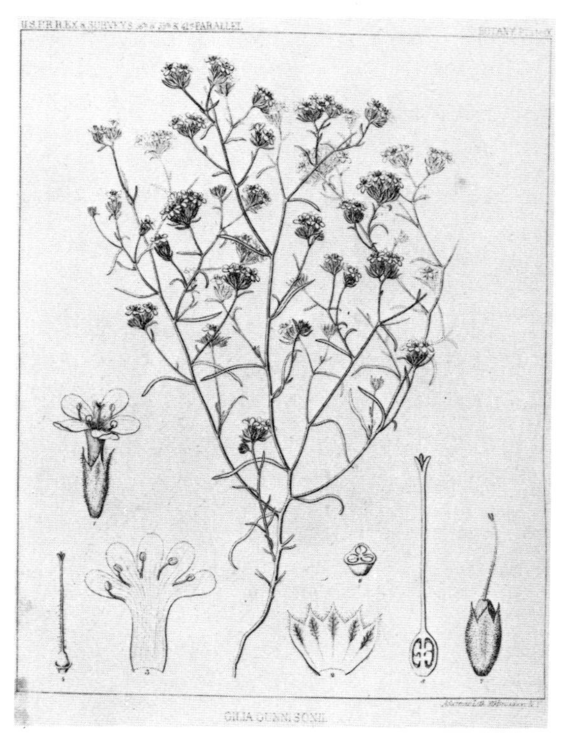

FIG. 23 "*Gilia Gunnisonii.*" (*Gilia gunnisonii.* Gilia.) "Ackerman Lith. 379 Broadway, NY." Lithograph; engraved on stone by Joseph Prestele; 29 × 22.5 cm. Plate IX of "Report on the Botany of the Expedition," by John Torrey and Asa Gray, in Lieutenant Edwin Griffin Beckwith's *Report of Explorations for a Route for the Pacific Railroad, on the Line of the Forty-First Parallel of North Latitude, 1854–55.* (Vol. II of the Pacific Railroad Survey Reports; Washington, 1855.) Courtesy the University of Delaware Library.

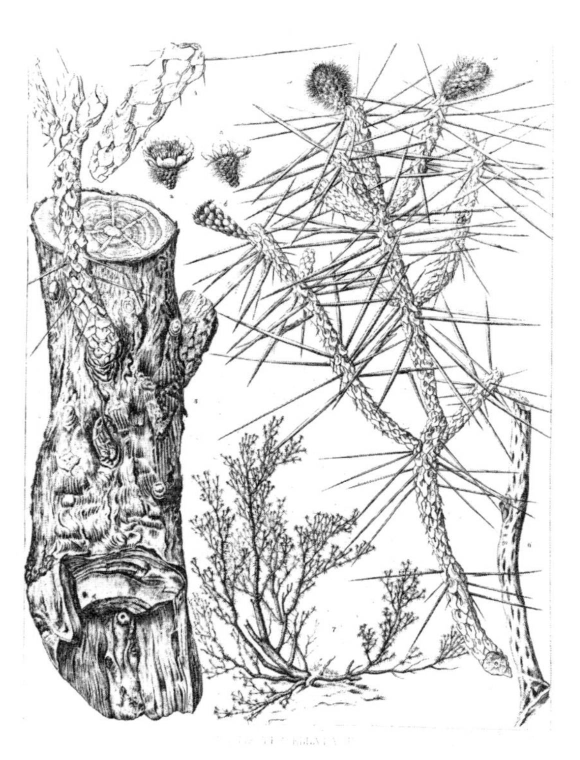

FIG. 24 "*Opuntia Tessellata, F.*" (*Opuntia ramosissima.*) Printed by "Ackerman Lith., 379 Broadway, NY." Lithograph; engraved on stone by Joseph Prestele; 29 × 22.5 cm. Plate XXI for "Description of the Cactaceae," by George Engelmann and J. M. Bigelow, in Lieutenant A. W. Whipple, *Exploration and Surveys for a Railroad Route from the Mississippi River to the Pacific Ocean . . . Near the Thirty-Fifth Parallel . . . in 1853–54.* (Vol. IV, Part V, No. 3, of the Pacific Railroad Survey Reports: Washington, 1856.) Courtesy of the University of Delaware Library.

revealing. Often—in this assignment as in others—before he could make the engravings, it was necessary for him to make the preliminary drawings from the dried specimens and rough field sketches sent him. Such alterations and corrections sometimes caused considerable trouble, for which he sought additional pay. The twenty-four plates of Cactaceae in Whipple's survey were particularly difficult. In sending proofs to Torrey on February 4, 1857, Prestele asked, "Do you not think $13 is a low price for such an engraving as Plate XVII & XVIII is. I can say I rather make 3 engravings of Mr. Sprague's Drawings than such a one. Dr. [George] Engelmann speaks in his letter to Capt. Whipple (which was sent in the package with the Cactacaes) that these Cactacaes could be done on steel in St. Louis for $120 a plate. All proofs that you will receive . . . will be more difficult

FIG. 25 "*Aster Bigelovii.*" (*Machaeranthera bigelovi.*) "Ackerman Lith. 379 Broadway, N.Y." Lithograph; engraved on stone by Joseph Prestele; 29 × 22.5 cm. Plate X for "Description of the General Botanical Collection," by John Torrey, in Lieutenant A. W. Whipple, *Exploration and Surveys for a Railroad Route from the Mississippi River to the Pacific Ocean . . . Near the Thirty-Fifth Parallel . . . in 1853–54.* (Vol. IV, Part V, No. 4, of the Pacific Railroad Surveys: Washington, 1856.) Courtesy of the University of Delaware Library.

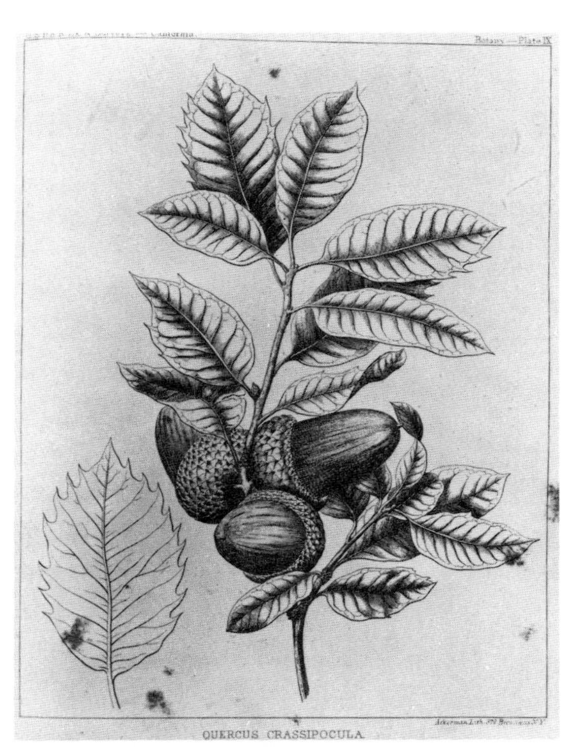

FIG. 26 "*Quercus Crassipocula.*" (*Quercus chrysolepis.* Canyon Live Oak. Goldencup Oak.) "Ackerman Lith. 379 Broadway, N.Y." Lithograph; engraved on stone by Joseph Prestele, drawing by E. Dwight Church; 29 × 22.5 cm. Plate IX for "Description of the Plants Collected Along the Route by W. P. Blake," by John Torrey, in Lieutenant R. S. Williamson, *Report of Explorations for the Pacific Railroad Upon Routes in California Connecting with Routes Near the Thirty-Fifth & Thirty-Second Parallels, 1853–54.* (Volume V, Article VII, of the Pacific Railroad Surveys: Washington, 1856.) Courtesy of the University of Delaware Library.

yet than all that I sent you already. You will please take it not amiss that I mention this."

Prestele was understandably anxious to see how his plates turned out. On December 4, 1857, months after they had been printed, he wrote Dr. George Engelmann, the St. Louis physician and botanist, asking for a set of the plates together with the text. Although Engelmann knew of Prestele, the artist apparently felt the need to introduce himself and to explain the reason for his request. He wrote: "I engrave[d] mostly the botanical Drawings for the Government at Washington, and also that I engraved The plants & Cactacaes for Lieut. Whipple's Report for which I suppose you have made the explanations for them, and also the one for Major Emory, which was I suppose done by your advice on steel at your City. As I take much interest with these works, . . . I would beg your kindness to let me see a proof of each of them, viz. those done on steel or if I dare ask your favor for [the] whole set with the letter-press, if so it would give me much pleasure." Prestele signed himself "engraver."[83]

Much of Prestele's work for the government was printed by the lithography firm of James Ackerman in New York, which had many government contracts during the 1850s.[84] Prestele's relations with Ackerman were generally harmonious. However, his relations with "the officers" in Washington, meaning the army officers involved with preparing the reports of the expeditions, were not always good. It is possible that the artist's language problem unwittingly created difficulties with strangers, and it is apparent from Prestele's letters that he was sometimes baffled and frustrated by the Washington bureaucracy.

Most frustrating of all were Prestele's difficulties in getting payment from the government. He had to submit two bills at the same time, each for half of his total charge. One bill went to the Clerk of the U.S. Senate, who paid promptly; the other to the Clerk of the House, who paid only after many months and repeated billings and letters. "So everything goes through trials and tribulations, misery and sorrow," Prestele philosophized, confessing to his friend, Christian Metz, that even after a lifetime of such experiences he had not "learned to believe and accept without grumbling and reluctance." He added that his difficulties with the government often reminded him of "the sad story with Vick in Rochester which had given me much pain and grief and worry."[85]

"The Sad Story With Vick"

WE DON'T KNOW THE DETAILS ABOUT PRESTELE'S "SAD STORY WITH VICK IN Rochester," which gave the artist such distress, but it may have involved illustrations in the *Horticulturist, A Journal of Rural Art and Rural Taste.*

In 1852, after the tragic death of Andrew Jackson Downing, the original publisher and editor of the *Horticulturist,* James Vick (1818–1882) acquired the magazine. Vick, a journalist in Rochester, New York, then moved the magazine to that city.[86]

Four years earlier, Prestele's engraving of the rose "Souvenir de la Malmaison" had been published in the *Horticulturist* (August 1848). Perhaps through that slight connection with the magazine or, as is more likely, because Prestele's work was known to Vick's editor, the urbane Patrick Barry (a partner in the Mount Hope Nursery in Rochester), Prestele was asked to supply illustrations for forthcoming issues. Vick would certainly have encouraged such a move for he, more than most publishers of his time in America, understood the popularity of illustrations, particularly colored ones which were then quite novel.

James Vick was a man of many interests, a tireless promotional genius with beguiling charm and more than a streak of idealism. He had been born near Portsmouth, England, and had been brought by his parents to America in 1833 at the age of fifteen. He learned the printer's trade in New York City and by 1850 was an editor of the Rochester, New York, *Genesee Farmer*, one of the leading agricultural journals of the day. His purchase of the *Horticulturist* followed. Vick was passionately fond of flowers and sensing that new postal regulations would encourage the mail-order selling of garden seed throughout the expanding nation, he started the James Vick seed firm in Rochester about 1856. Within a few years his innovations in that generally staid business led one writer to comment in some despair, "No one knows what he will do next." He may have been the first to use colored plates in his catalogues, which he filled with advice on all phases of gardening, including the cultivation of plants, landscaping, and much else, all presented in such a chatty, informal way that "Mr. Vick" became an oft-quoted celebrity throughout rural America. Realizing that efficient and inexpensive postal service was essential to the growth of his business, he carried on a constant and vehement battle with the postal department, all of which he vividly reported in his catalogues, and which was followed by fascinated readers. Along with the skillful promotion of both his business and himself, Vick sold good seed at fair prices.

In announcing his purchase of the *Horticulturist*, Vick described the changes he planned to make. Among other things he promised a full-page engraving of "some new, rare, and valuable fruit or flower, drawn from nature, and engraved in a style not excelled," in each number. These plates, he asserted, would "add to the value of the work, and meet the improving taste and increasing wants of the horticultural community." Subscribers had the choice of an edition of colored plates at $4 a year, paid in advance, or uncolored plates at $2 a year.[87]

Vick obtained his engravings from many different sources. Some had already been published elsewhere, and most of these have the signatures of their artists and engravers. Vick also published some twenty-four plates engraved on stone and four drawn in "chalk" on stone which can be attributed to Joseph Prestele on the basis of the lithographic techniques used, the design style, and the captions. The captions are particularly revealing; most have the individualistic lettering which can be identified from other sources as by Joseph Prestele (see Fig. 27), and a few are elegantly engraved in a script that closely resembles the artist's own handwriting. (See Plates 39, 42.)

Most of Prestele's illustrations in the *Horticulturist* are of fruit, particularly pears, then immensely popular with growers; a few are of flowers. The first of these illustrations appeared in the January 1853 issue and continued to be used

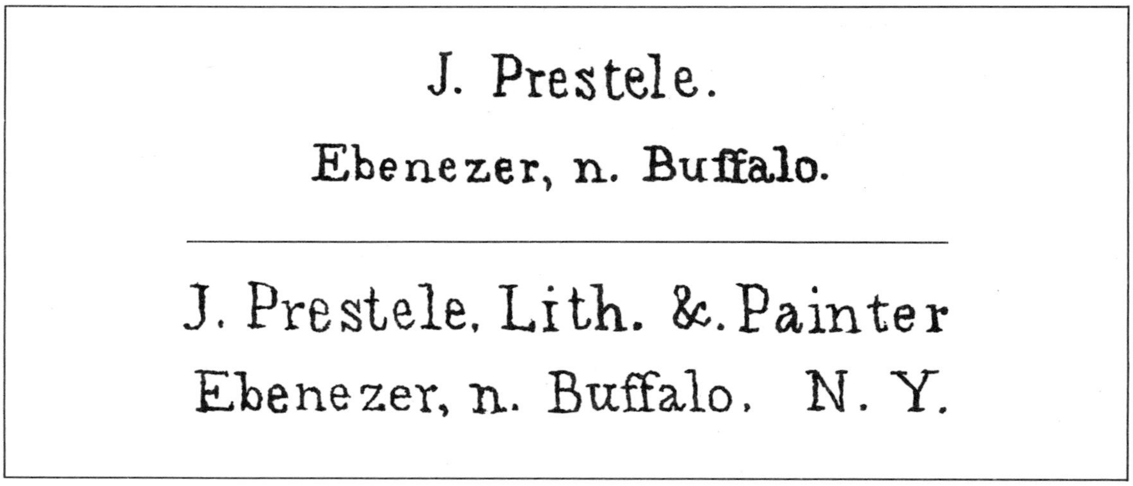

J. Prestele.

Ebenezer, n. Buffalo.

J. Prestele. Lith. &. Painter

Ebenezer, n. Buffalo. N. Y.

FIG. 27 Hand-lettered signatures of Joseph Prestele on nurserymen's plates, before 1858. Author's collection.

The distinctive type of lettering used by both Joseph Prestele and his son Gottlieb are helpful in identifying their work. Here are examples used on much of Joseph's work while he was living at Ebenezer, New York, and perhaps during his first years at Amana, Iowa. Such lettering appears on both his nurserymen's plates and his illustrations in the *Horticulturist*.

Joseph confessed to Asa Gray that he did not feel competent in the art of hand lettering, and these examples make clear his limitations.

nearly every month until 1855 when Vick sold the magazine.

Since so many of Prestele's plates were actually published in the *Horticulturist*, perhaps the "sad story" he referred to concerned his compensation by Vick who was no doubt more than a match for the gentle Prestele.

Prestele's association with the magazine petered out under the aegis of the new publisher, Lloyd Pearsall Smith (1822–1886), an editor and librarian. Smith moved the magazine to Philadelphia where it was issued from July 1855 until it passed to a new owner in June 1858. The editor was Smith's father, John Jay Smith (1798–1881), a well-known author and editor of various botanical works and, like his son, a librarian. During the Philadelphia years, at least three plates by Joseph Prestele were used—perhaps left over from the Vick regime: the "Red Astrachan" apple (October 1855), "Knight's Early Black" cherry (October 1856), and "Beurre Superfin Pear" (January 1857).

Plates for Nurserymen

ONE OF THE MOST IMPORTANT AND LEAST-KNOWN ASPECTS OF JOSEPH PRESTELE'S career was the part he played in the development of a specialized type of art, the colored fruit and flower plates made to aid nurserymen and their agents—the "tree pedlars" who traveled about rural America and Canada—sell plants.* That development came well before colored illustrations began to be used in seed and nursery catalogues; indeed, the success of the nurserymen's plates may have suggested the use of color illustrations in catalogues and for other commercial purposes.

Prestele was the first to make and sell these plates, as considerable evidence shows, and as his youngest son, William Henry, claimed in the mid-1870s. According to him, his father had "*'ORIGINATED' and made the first Lithographed Fruit and Flower Plates*, for the use of Nurserymen in the U.S." during the late 1840s.[88] The idea of making plates for nurserymen was only one step removed from the fruit and flower illustrations Joseph Prestele had made during 1848–49 and later for the *Horticulturist*, and the *Transactions* of the Massachusetts Horticultural Society, and, indeed, some of his nurserymen's plates were identical to earlier designs or were only somewhat revised. Some of his plate designs can be dated to before 1852.[89]

By December 1857 Prestele had on hand plates of more than "100 sorts," as he wrote Dr. George Engelmann in St. Louis, explaining that along with his scientific illustrating, he was "also Fruit & Flower Painter such which are used by Nurserymen." He mentioned that the St. Louis nurseryman, John Sigerson, was one of his customers.[90]

For his nurserymen's plates, Prestele adapted the art of botanical illustrating to the commercial needs of nursery selling, principally by eliminating details important to botanists—such as seeds, roots, and cross sections of blossoms—but which nurserymen did not want or need. The results he achieved were similar to the best illustrations in horticultural works of the period, in which the fruit or blossom shown was emphasized, and the composition planned to be visually arresting. But Prestele's work was almost too good for his market. He would not compromise on quality, and furthermore he did not promote sales, both because that wasn't his nature, and because it is unlikely that the tenets of his church would have approved advertising or other such worldly activities. He issued plates of at least 250 subjects, most of them of fruits and berries, but neither the size of his production over the years nor the amount of his sales can be determined. We know that his prices were competitive despite his higher production costs; it seems likely that his customers were relatively limited in number.

One of Prestele's best customers was the Mount Hope Nursery at Rochester, New York, which was rapidly becoming the largest and most respected of its kind in America and was helping to make that city a center of the nursery

---

** A checklist of nurserymen's plates, signed by or attributed to the Presteles, is given in Appendix C.*

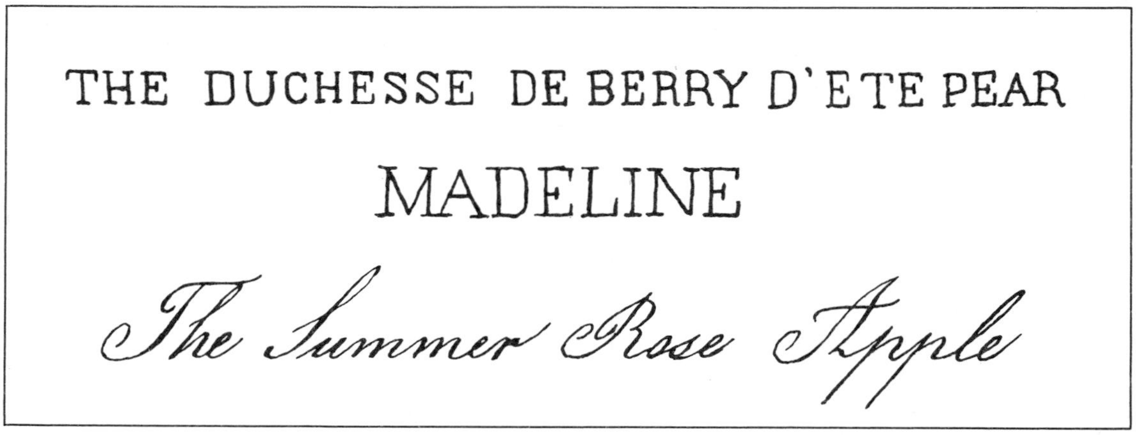

THE DUCHESSE DE BERRY D'ETE PEAR

MADELINE

The Summer Rose Apple

FIG. 28 Typical lettering in captions of nurserymen's plates by Joseph Prestele, and Gottlieb Prestele. Reproduced from examples in the author's collection.

The two types of block letters shown above—both on nurserymen's plates signed "G. Prestele"—were used on many plates produced at Amana, Iowa, and are believed to be characteristic of Gottlieb's lettering. The handwritten caption (below) could be either by Joseph or Gottlieb. Written captions are found on what appear to be either the original watercolor designs for plates, or proofs of them before lettered captions were added. Examples of handwriting by the two men are shown in Figs. 17, 32, and 37.

trade.[91] The nursery had been founded in 1840 by two partners, one, a German immigrant named George Ellwanger who was highly trained as a nurseryman, and the other, an Irish immigrant of good education and background named Patrick Barry, the same Barry who was later to become editor of the *Horticulturist*. (Besides being an editor, journalist, and author, Barry was an able administrator and a lecturer on horticultural subjects.) Indeed, Prestele's sales to the firm began in April 1854, one year after he had begun to do illustrations for Barry at the *Horticulturist*.

Through the succeeding years their purchases continued and a friendship of sorts apparently developed between the artist and the proprietors.[92] Perhaps Barry or Ellwanger checked a proof copy of the artist's "Summer Rose Apple," for penciled on the plate is: "The yellow is too dark. Should be lighter."[93] Misrepresentation was anathema to the firm as, indeed, it was to the artist. Years later, H. B. Ellwanger, son of George, commented that the firm viewed with distaste "exaggerated color plates."[94]

From April 22, 1854, until March 23, 1860, the firm bought plates from the artist, the largest orders being in 1856 and 1857. The purchases during the five years totaled some $985.76 which, if the firm was charged the stated price of 25 cents for the fruit prints and 37½ cents for flower prints, represented between

2,000 and 3,500 plates.[95] The nursery issued bound assortments of these plates to their agents, to nurserymen, and perhaps to others into the 1870s.[96]

OVER THE YEARS GOTTLIEB PRESTELE, THE PATIENT STUDENT, ABSORBED HIS father's skills and gradually took over more and more of the work. (His elder brother Joseph, Jr., had left home in the winter of 1843–44; his younger brother Willim Henry departed before 1858.) Gottlieb, like his father, signed his plates only infrequently. His first nurserymen's plates were done at Ebenezer and are signed "Lith. & col$^{\underline{d}}$ by G. Prestele. Ebenezer, near Buffalo N.Y." Another signature, "G. Prestele," may also date from his residence there. The father's signatures at Ebenezer include "J. Prestele, Ebenezer, n. Buffalo"; "J. Prestele, Lith. & Painter, Ebenezer, n. Buffalo, N.Y."; or simply "J. Prestele." The order in which Gottlieb and his father used those signatures is difficult to determine. (Other signatures were discussed on p. 73.)

During the 1850s the nurserymen's plates created by Joseph Prestele and his son Gottlieb began to attract increasing attention although apparently, as noted previously, the Presteles never advertised or otherwise promoted them. The traveling agents of nurseries found the plates very useful because they were attention-catching at a time when colored pictures of any kind were a novelty, and they faithfully pictured their subjects. Horticultural and agricultural journals paid Prestele the dubious honor of "borrowing" some of his designs to use as illustrations without giving him credit. Some of these copies or adaptations were poorly done black-and-white wood engravings.

More serious was the competition that began to develop when aggressive and enterprising publishers and lithographers—and several speculators—took up the colored plate business.[97] Among these were the Hartford, Connecticut, lithographers, E. B. & E. C. Kellogg,[98] who before 1859 had begun issuing their "Kellogg's Series of Fruits, Flowers, and Ornamental Trees." A comparison of their plates with those of Prestele's show that some of their designs are identical to those known to have been published by Joseph, even being the exact size; some are reverse images. None credit the artist. It is not known what arrangement the Kelloggs made with Prestele, or how they acquired Prestele's designs, but nothing has been found to suggest that they simply "borrowed" them.[99] The Kellogg Series was only a small part of their large print business and, judging from the few examples located, never posed a real threat to the Presteles' sales.

The real competitor, and the one whose relations with the Presteles were murky, was Dellon Marcus Dewey (1819–1889), a Rochester bookseller, publisher, dealer in art, and an amateur architect who delighted in horticulture. With

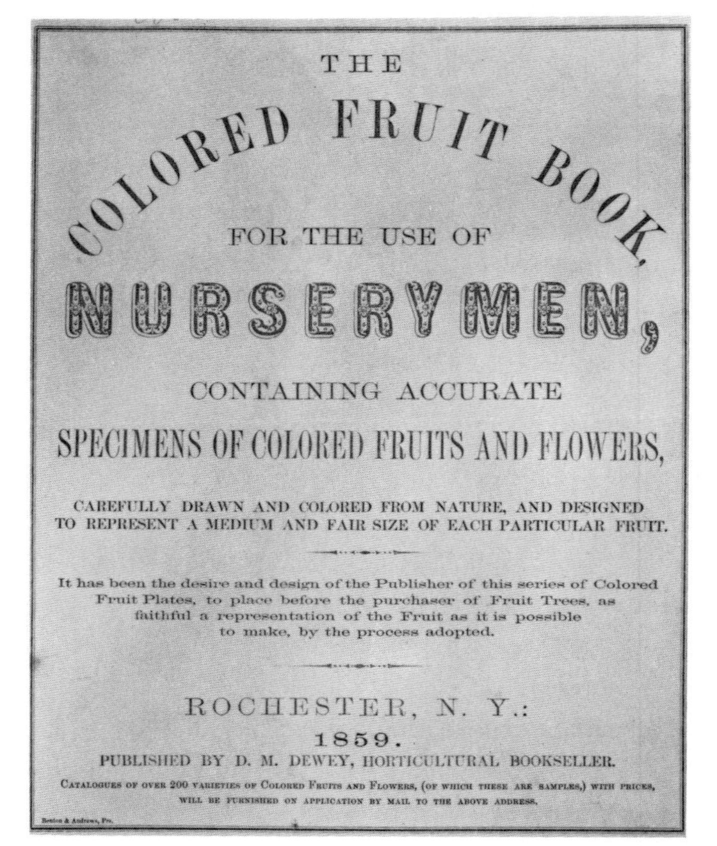

THE

COLORED FRUIT BOOK,

FOR THE USE OF

NURSERYMEN,

CONTAINING ACCURATE

SPECIMENS OF COLORED FRUITS AND FLOWERS,

CAREFULLY DRAWN AND COLORED FROM NATURE, AND DESIGNED TO REPRESENT A MEDIUM AND FAIR SIZE OF EACH PARTICULAR FRUIT.

It has been the desire and design of the Publisher of this series of Colored Fruit Plates, to place before the purchaser of Fruit Trees, as faithful a representation of the Fruit as it is possible to make, by the process adopted.

ROCHESTER, N. Y.:

1859.

PUBLISHED BY D. M. DEWEY, HORTICULTURAL BOOKSELLER.

CATALOGUES OF OVER 200 VARIETIES OF COLORED FRUITS AND FLOWERS, (OF WHICH THESE ARE SAMPLES,) WITH PRICES, WILL BE FURNISHED ON APPLICATION BY MAIL TO THE ABOVE ADDRESS.

FIG. 29 Title page of D. M. Dewey's *Colored Fruit Book*, 1859. Author's collection.

It is not clear whether Dewey sold these books, or gave them to nurserymen and their agents to promote his venture. The first owner of this volume was "E. B. Lamborn" whose name is stamped on the cover; he has not been identified.

these interests, and his location in such an important nursery center as Rochester, it is not surprising that he took up Prestele's good idea and threw all his energies and promotional abilities into the production and sale of nurserymen's plates.

Dewey launched his new venture in 1859 with an initial stock of plates largely acquired from others, such as the Kelloggs in Hartford and Joseph Prestele; and also with reprints of Prestele's illustrations in the *Horticulturist*. He had the Buffalo lithographers J. Sage and Sons[100] make a number of plates of ornamental trees, and arranged for inexpensive prints to be made from drawings by Anton Hochstein, a young German immigrant whom Prestele knew.[101] These were engraved on wood and printed in green ink by George Frauenberger, another German immigrant who set up shop in Rochester in the late 1850s.[102] Both Hochstein and Frauenberger were good artists, but Dewey only wanted them to supply prints that could be sold more cheaply than his regular line; the results were poor.

Dewey issued his first list of colored plates early in 1859, followed by a second a few months later after he had added twenty-two more titles, mostly of ornamental trees and flowers.[103] To further promote his venture, Dewey produced

FIG. 30 Frontispiece from D. M. Dewey's 1878 *Classified Catalogue of Original Colored Fruit Plates, No. 4.* Author's collection.

This colored frontispiece was intended to advertise the quality of Dewey's "Engraving, Lithographing and Chromo Color Printing," all used in the production of his nurserymen's plates. He said that he then had a stock of "over 2,000 varieties of fruits, flowers, ornamental trees, &c." and "nurserymen's requisites."

The Colored Fruit Book, for the Use of Nurserymen, which consisted of a printed title page, seventy representative prints made from his initial collection including some by Prestele, and a copy of his second list of plates with prices. (We don't know if Dewey paid Prestele for these or not.) Dewey stated that his plates had been carefully drawn and colored from nature, and that their purpose was to "place before the purchaser of Fruit Trees, as faithful a representation of the Fruit as is possible to make, by the process adopted."[104] No matter where he had obtained the plates Dewey unabashedly imprinted them with "D. M. Dewey's Series of Fruits, Flowers, and Ornamental Trees." Not only did he claim the plates as his own, but he also took the title which the Kelloggs had previously adopted for their line.

Dewey later recalled that his enterprise was "hailed with great favor and received the immediate patronage of the nurserymen throughout the country. . . . It soon became necessary to enlarge the catalogue by adding to it the popular varieties . . . grown in all sections of the country."[105]

Having "tested the wind" with encouraging results, Dewey entered the field in earnest. His first need was an inexpensive assembly-line method of pro-

FIG. 31 "*Purple Wistaria*." Unsigned watercolor; produced and published by D. M. Dewey, Rochester, New York; 27 × 21 cm. From a collection of nurserymen's plates, bound and sold by the Bloomington Nursery, ca 1869–75. Author's collection.

This is an example of the freehand work developed by Dewey and perfected by his artists and colorists. At the same time, they were mass producing plates by Dewey's theorem process, in which stencils were used to guide the work of the colorists.

ducing large numbers of plates at a low cost. His knowledge of art proved helpful here, and after some experimentation, he chose a variation of the old theorem process by which the main outlines of a design were laid on with stencils and the details added freehand.[106]

He proceeded to either copy directly or simplify designs by the Presteles. As his colorists became more skilled, they sometimes made plates almost entirely freehand, in handsome, almost abstract designs.[107] For a less expensive line he used wood-engraved and chromolithographed plates in addition to those routine designs made primarily with stencils.

It was impossible for Dewey to keep secret the theorem process he had developed for making his popular "hand-painted" plates, and it was equally impossible for him to retain in his employ all those whom he had trained in that work. A number of the defectors set up their own plate businesses causing Dewey to comment that "females and others" had attempted to copy his plates but, he added with satisfaction, nearly all of them had failed.[108]

Dewey didn't try to imitate the quality of the Presteles' work. Clever as he was he must have sensed that the mass market he was trying to reach didn't require Prestele-quality plates. This audience preferred reasonably accurate portrayals of fruits and flowers presented in rich, eye-catching colors, and moderately priced. He was also responsive to changing times and changing needs, something that the Presteles never were. Dewey expanded his line to include plates in a number of sizes.[109] He also wrote and published an astonishing *Tree Agents' Private Guide* that included the advice that these salesmen should impress upon whomever they should meet on their travels that they were upright, moral, and God-fearing men; it also had instructions on how to pronounce the French names of plants; to dress in neat and clean clothes—and how to clinch a sale.[110]

Dewey finally retired in 1888, the year before his death. At that time he announced—with pride of achievement and disregard of evidence—that he was the "inventor and originator of Color Plates," and that he had produced the "first plate ever made."[111]

Although Dewey's claim was unfounded, the matter of priority was not as important as the real contributions both he and the Presteles made to the specialized art of nurserymen's plates. Joseph Prestele created plates of great beauty in the fine art tradition, but his influence was limited; he was an artist, not a promoter. Dewey used several different techniques, including his creative and distinctive hand-painted work, to produce a popular art of more than average quality. This tireless man with his fertile, inventive mind, exploited to its full potential the nurserymen's plate business and dominated it throughout its best years.[112]

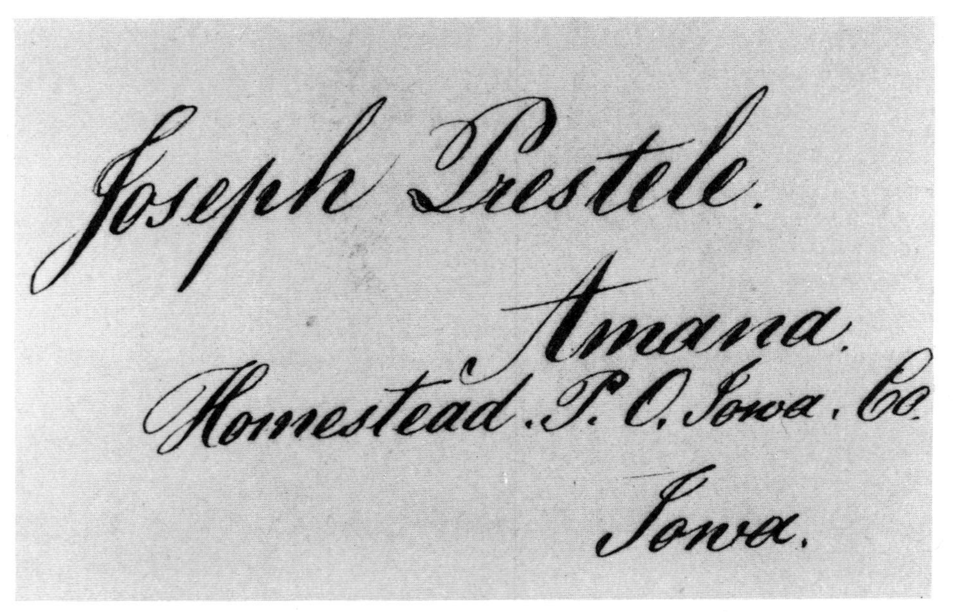

Joseph Prestele.
Amana.
Homestead. P. O. Iowa. Co.
Iowa.

FIG. 32 On September 8, 1858, before his departure from Ebenezer, New York, to the Inspirationists' new settlement in Iowa, Joseph Prestele sent Asa Gray this handwritten card (5 × 8 cm.) giving his new address. Courtesy, Gray Herbarium of Harvard University.

3 Iowa, the Amana Settlement

Prestele's Last Years, 1858–1867

WITHIN A DECADE AFTER THE INSPIRATIONISTS HAD ESTABLISHED THEMselves at Ebenezer, the outer world, with all that implied, began crowding in upon them. The allurements of Buffalo, and the infectious excitement created by the countless families moving west along the Erie Canal, both only a short distance away, made the younger members restless. Most disrupting of all, and in unsettling contrast to the well-established order of European life, was the freedom offered by wilderness America, with its accent upon the individual. Now, although great labor and much of the Community's resources had been invested in transforming an "Indian Wilderness" into the orderly agricultural-industrial settlement of Ebenezer, it was becoming apparent that the Inspirationists could not remain there and hope to continue in their close-knit religious and conservative ways.

In 1854 Christian Metz, still one of the church leaders, had another message from the Lord instructing the members to "direct your eyes toward a distant goal in the West to find and obtain there a start and entrance or a settlement."[1] In obedience to that directive a search committee visited Kansas and other locations in the Middle West, finally reaching the fertile valley of the Iowa River in the state of Iowa which had been admitted to the Union in 1846 but as yet was only sparsely settled.

The search committee noted with approval the meandering river and lush meadows along its banks, covered with tall prairie grasses constantly stirring in the wind, and made bright by wildflowers whose colors changed with the seasons. The soil, they found, was deep, black, and unbelievably fertile. On the low hills bordering the valley was a forest of hardwoods, one of the few in southeastern Iowa. Elsewhere, clumps of ancient trees, like dark, green islands, rose above the meadows. The land was rich with promise and had the isolation the elders sought. A tract of 18,000 acres was purchased, later increased to 26,000 acres. They named their new home Amana, meaning "Believe faithfully."

The Community's leaders planned the move west carefully to ensure both that their Ebenezer property was disposed of to the best advantage and that houses would be available when the settlers arrived in Iowa. The orderly transfer extended over ten years.

Prestele had written Gray in 1855 of the plan to resettle in Iowa, but the artist's removal was not ordered until the fall of 1858. On September 8 of that year he informed Gray that he was to depart in about four weeks. "I am very sorry to come now so far from each other," Prestele wrote, "which causes that I can not do any work more for you, but will be remembering allways your kind favors you bestowed to me, since our acquaintance to annother and yet I shall remain in your friendship." He enclosed a card on which he had written his new address.

Prestele must have found the move west physically as well as emotionally difficult. He was sixty-two, he had never been strong, and he was going far from the associations that had meant so much to him. In the record of his life there is little evidence of joy, but much of resignation. How he and his family and their companions traveled to their new home has not been learned. Perhaps they went as did the first party to leave Ebenezer. That caravan headed west in July 1855, the covered wagons filled with household gear, tools, some machinery, and other equipment needed to open the new settlement. Livestock was taken along too. Details of the route taken by that first party are lacking, but it is reported that they went by lake vessel from Buffalo to Chicago, then by rail to Rock Island on the Mississippi, and by steamboat to Muscatine, Iowa. From there they headed west across rolling countryside to Homestead, a village adjoining their tract (and which they later purchased).[2]

The journey must have been an exhausting experience for Prestele. In addition to the confusion and strain of travel under conditions difficult at best, there were household goods to look after, and his cumbersome lithographic press. We know, as well, that he also brought along many lithographic stones, both with and without designs, since many of the nurserymen's plates that Joseph and Gottlieb had produced at Ebenezer were reissued at Amana. Those brittle stones would have had to have been very carefully crated, particularly those with images upon them. At each of the different stages in their journey those heavy crates would have had to be loaded and unloaded.

At the new settlement, seven villages gradually took form. Scattered along the north ridge of the valley were East Amana, Amana, Middle Amana, High Amana, and West Amana. South across the Iowa River, on the ridge opposite, were South Amana, and the town of Homestead. The villages were so spaced that farmers living in them were conveniently near the fields assigned to their care. No farmhouses were built outside of the villages because, among other reasons, church services were frequently held, and all meals were taken in communal dining rooms.

In time, as the wide valley along the Iowa River came under cultivation, the 26,000 acres bought by the Inspirationists began to resemble a great park. Its river, fields, and woodlands were left without fencing except where cattle were pastured; the margins of roadways were kept free of brush. On the low hills above the valley fields, the villages with their modest buildings of unaffected Old World designs provided accents in the natural scene, suggesting prosperity and the continuity of tradition. But it was many years before that happy scene was created; much hard work had first to be invested.

In all the villages, the houses were generally of two-and-a-half stories, their design and construction essentially Germanic but modified by the American experience of the settlers. Some buildings had sturdy frames and were weatherboarded, their unpainted exteriors soon bleached to a somber gray. Here, unlike Ebenezer where the frame buildings were painted, the Society decided that it was more economical to rebuild when the structures deteriorated, than to preserve them with paint.[3] Other houses were made of the rich brown sandstone found on the site. (A few of these had date stones set in the gables giving the initials of the builder and the year of construction. Such date stones were a Germanic tradition but their use here in Amana hints of forbidden vanities.) Some houses were of brick, often with small windows under the eaves to provide added light and ventilation to attic rooms. The principal buildings were the mills and factories, and the churches—one of which was in each village. In Amana the church was a long, sandstone building whose austere interior was furnished with stoutly made wooden benches, splay-legged and straight-backed.

Each village was patterned in the old-fashioned German manner, with a cluster of from forty to a hundred houses crowding upon a long, straggling lane with several irregular offshoots.[4] At Amana, which became the main village, the commercial and manufacturing operations included a woolen mill and a factory for printing calico. Waterpower was provided by a canal some six miles long which was dug from the Iowa River above the village, through the valley, to the plants. It was a herculean feat, begun in 1865, and accomplished by men with hand shovels and primitive ox-drawn scoops.

At the new settlement, as at their previous homes, life was ordered by the tenets of the sect and the inspirational revelations of the church leaders. Much was forbidden—but there were exceptions. Celibacy, for example, was a matter of church doctrine, but there were many children and on these were lavished the delights of childhood everywhere. They had an abundance of toys as well as little benches, chairs, and other furniture made by village craftsmen and scaled to the

FIG. 33 Wood engraving of Amana, Iowa, about 1870. (Charles Nordhoff, *The Communistic Societies of the United States*, Harper & Brothers, 1873.)

Less than twenty years after the first settlers had begun construction, Amana had become a thriving community and the principal town among the seven Amana villages. Seen here is a view of the meandering main street, lined with houses. In the background is the mill and a factory. The European style of dress worn by members of the Community, examples of which are shown here, was established by the church leaders. Women generally wore white caps, kerchiefs, and aprons, with dresses in dark colors.

size and pleasure of little people. While any expression of individuality was expressly prohibited—for example, of fashion and vanity in dress—the domestic arts of knitting, crocheting, quilt making, and other activities whose products must have brightened homes flourished. The furniture made at Amana for their homes was constructed of maple, walnut, and other native woods with exceptional nicety and in designs popular in early nineteenth-century central Europe. They were simplified, it is true, but not to the point of austerity. Ornamental trees and shrubs were considered a mark of worldliness and could not be planted to relieve the drabness of unpainted buildings and fences. But at least one ancient oak was allowed to remain on the site of Amana, and fruit trees and grapes were permitted, the latter being grown on trellises placed against the houses. Paradoxically too, flowers, including the bright red Flanders poppy, were not denounced as a "pleasure to the eye," but were grown everywhere, "around each dwelling, in front of the church, and even in the hotel and school yards."[5] And the joyful observance of Christmas customs included decorated Christmas trees.

Assignments for all work were made by the elders who, as in the past, even chose the trade or profession each member of the church was permitted to follow. The women, for example, were assigned the task of cultivating the communal vegetable gardens near the villages. The basic assumption was that each member would contribute by his or her labor to the welfare of the Community and that each would do so willingly, but there was no way to enforce the labor rules other than by admonition and social pressure. These were not always effective and, as time passed, more and more "drones"—as the Community called them—found ways to live on the industry of others, and this was one of the causes of the final dissolution of the religious community's formal organization in 1932.

During the first years of settlement when the villages were being created, there was much confusion and the messiness of the unfinished and the not-yet-begun; of lanes muddy and dusty in turn, noisy with the movement of animals; of smells, flies, summer heat, winter chill. Much of the life there was similar to what the members had known in Ebenezer, but the newness, the confusion, the never-ending task of transforming the Iowa countryside into a controlled agricultural-industrial community must have been difficult for Joseph Prestele and his family. Not the least among the adjustments required of the artist came from the change in the pattern of his work and of his association with learned men and events, for his career as an illustrator of scientific works appears to have ended with his move to Iowa. He was—as he had mentioned to Gray—too far from potential commissions; the frontier had no need for his services, and the Civil War soon came to distract funds and interests from peacetime projects. Lacking commissions for the work which had given him his greatest pleasure, Prestele turned full time to making nurserymen's plates as soon as circumstances at his new home permittted. Gottlieb, the middle son, remained his father's faithful assistant.

The Presteles may have expected that their sales would be limited to the nurseries then being established throughout the new West, but Ellwanger & Barry and other old customers in the East continued to buy their plates until the market for such unessential items shrank soon after the outbreak of the Civil War. Nevertheless, father and son patiently continued their platemaking. Modest recognition of their artistry came, but probably only limited orders.

After the Society's move to Iowa, there are hints of increasing restraints to individual expression. The names of Joseph and Gottlieb disappeared from their prints, and only the Amana Society's imprint was used.[6] That change was also reflected in the Ellwanger & Barry account books. Although Prestele's income had always gone directly into the Society's coffers, the firm's accounts had been with "Joseph Prestele." However, on March 21, 1859, shortly after he arrived in Amana, a new account was opened with the "Amana Society."[7] When the *Muscatine* (Iowa) *Weekly Journal* published a story about Prestele's nurserymen's plates, his name wasn't even mentioned. Rather the paper commented that the "Amana Society's . . . colored lithograph fruit and flower plates have become celebrated for their beauty and truthfulness. The minutest details are not forgotten, and the fruit stands out from the picture almost ready to grasp. As an evidence of the appreciation of their work . . . the plates of the Agricultural Department, at Washington, are all coloured by them. They also do the lithographing for many of the eastern nurseries."[8]

About 1860, some two years after he arrived in Amana, Joseph Prestele (or perhaps the Amana Society in his behalf) issued a list of plates available from his inventory entitled *List of Fruit—and Flower—Plates. Drawn from Nature. Lithographed and Colored by Joseph Prestele Sen. Amana Society, Homestead, P.O., Iowa Co., Iowa.*[9] (See Illus., p. 330.) It names 232 plates divided into the categories of "Apples, Siberian Crab Apples, Pears, Plums, Cherries, Peaches, Strawberries, Grapes, Currants, Miscellaneous [berries and nuts], Raspberries, Gooseberries, Flowers & Shrubs," and "Roses." The fruit and flower plates were priced at 25 cents each; the roses and grapes at 37½ cents—the same prices charged by Dewey in 1859. Although only abbreviated captions were used, and the errors in nomenclature and spelling were frequent, the list provides additional evidence of the many plates that were printed by Joseph and/or Gottlieb without signatures.

Included among the subjects listed is some of Joseph Prestele's last work, plates created amidst the unsettled conditions of the new community and during a time of Prestele's increasing ill health, his tuberculosis worsening. Nevertheless, these plates are among his finest art. One was an untitled print apparently designed for framing, of a bouquet of flowers. (See Plate 84.) His new plates for nurserymen included roses, ornamental vines and shrubs, and fruits developed and promoted by growers in the upper Midwest. Often the fruits are shown approximately life size—somewhat larger than most of his earlier designs. The coloring is rich but not unnaturally so; the "bloom" on plums and grapes is remarkably delicate. (Apparently he or Gottlieb colored the plates only when orders were received, as uncolored ones exist in Amana collections.)

Also, unlike his earlier designs for nurserymen's plates, he often included cross sections of the fruit, particularly apples, to aid in distinguishing varieties. Some compositions were far more detailed than the usual nurserymen's plate—such as that of the "Common European Walnut," which includes a leaf, a blossom, and green and ripe fruits, shown both whole and in cross section, details that other publishers of such plates would never have illustrated.

Often Joseph and Gottlieb drew and painted for their own pleasure and made gifts for the pleasure of others. These are among their most imaginative works. One is an uncolored lithograph, meticulously engraved on stone by Joseph, with vignettes of the places where the Inspirationists had lived during their

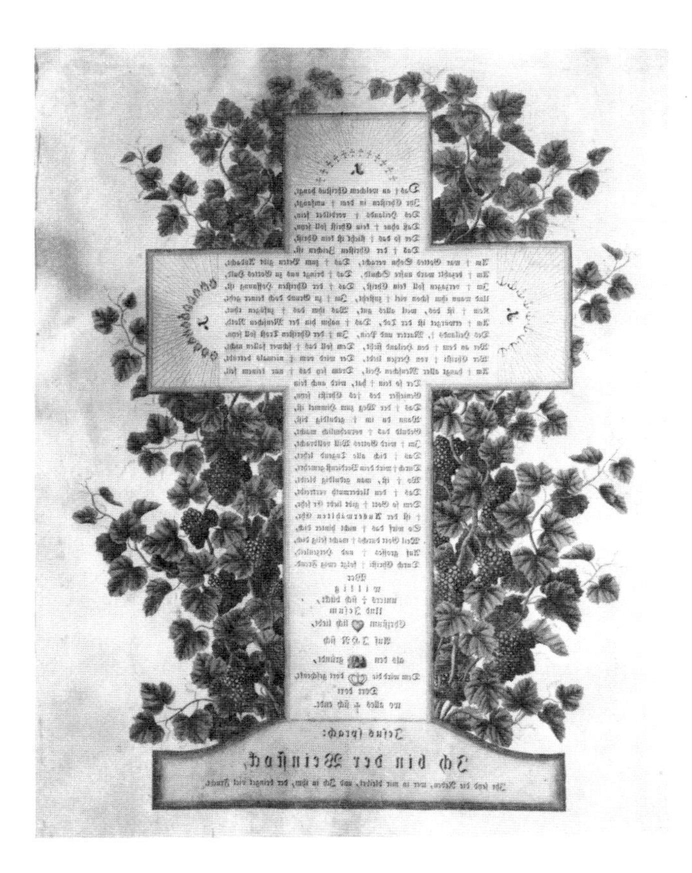

FIG. 34 Religious Print. Unsigned lithograph, attributed to Joseph Prestele, Sr.; engraved on stone, colored; 44.5 × 34.5 cm.

An older version of this print, in the Museum of Amana History, was probably brought from Germany by the Inspirationists and was the prototype for this print. Several copies of this later plate are in Amana collections, reinforcing the belief that they were made by the Presteles.

Every detail of the print has a symbolic religious meaning. The verse can be translated as: *He who bends under the (Cross) willingly and loves Jesus Christ (heart)ily Is given the (crown) When he is there where all (Cross) is ended.*

The general title at the bottom of the Cross refers to a biblical quotation: Jesus said, "I am the vine, ye are the branches. He that abideth in me, and I in him, the same bringeth forth much fruit; for without me ye can do nothing." (John 15:5, King James Version.) Courtesy, Mrs. Lina Unglenk, Amana, Iowa. Translation by Elisabeth Kottenhahn.

wanderings in central Europe and, in the center, a sailing ship riding the waves to New York City. This was the preliminary version of a design later completed by the addition at the bottom of several views of their new villages at Ebenezer, New York. (See Fig. 11.)

The Presteles seem to have taken particular pleasure in creating unusual pieces, often as gifts for friends. One imaginative and charming design by Joseph shows a delicately shaded blue vase containing a graceful arrangement of strawberry vines with fruit (Plate 86). It is a handsome composition but improbable because no strawberry vines could have been arranged in such a manner. Another time, on the occasion of four couples being married in a joint ceremony, Joseph painted a memento for each couple consisting of a rose, a poem, and an inscription, "In blessed memory on the morning of June 29th, 1860."

Of all the art created by Joseph and Gottlieb, there are several pieces preserved in Amana collections that express most clearly the depth of their religious feelings. One is a hand-colored lithograph of a cross whose outline frames a religious quotation in German. The design, which includes symbolic grape vines luxuriant with leaves and purple fruit as a background for the cross, was obviously copied from an older print, possibly brought from Germany and now in the Museum of Amana History. Another of these religious paintings, by Gottlieb, has an intricate design consisting of many small details—flowers, cherubim, fruit, ribbons, and other elements—minutely painted in various colors, suggesting half-remembered baroque folk art of Bavaria, and carefully planned symbolism. (See Plate 87.)

FROM THE AUTUMN OF 1858, WHEN JOSEPH PRESTELE AND HIS FAMILY LEFT western New York for Iowa, until the last tense months of the Civil War seven years later, there was no correspondence between the artist and Gray. Then, by a strange coincidence, a letter from Gray arrived at Amana on February 13, 1865, just as Prestele was finishing a letter to him. It "looks like Providence having brought us together anew, perhaps for some good purpose," the artist commented. He answered Gray's query about the whereabouts of a "very small" lithographic stone on which he had engraved the *Robinia viscosa* and *Cerasus virginiana* plates for the tree book by saying that he had sent it to Boston (to Tappan & Bradford?) in September 1858. He closed his letter with the sad comment:

> *Out here I do not prosper so much as might be expected and I desire and I wish you had work for me to do like in days past which were agreeable times to me.*
>
> *Very respectfully*
> *Your most Obt. Servt.*
> *Jos. Prestele*

Apparently there was no further exchange between the old friends. As the months passed the artist's incipient tuberculosis worsened. He had always been frail; the physical hardships he had experienced and the sedentary nature of his work combined to weaken him, and he was growing old. His last illness was described in the Community's *Inspirations-Historie.*

> *Since he hardly could eat he slowly became emaciated and it was visible that his life would soon end. Still, in his heart he had a glimmer of hope that perhaps with the coming of spring he would begin to feel better, which hope . . . he gave too much room in his heart, and so it happened that he was overtaken by death quite by surprise . . . March 9th., 1867.*[10]

His widow, Karolina Russ Prestele, survived him by some three years. She died January 6, 1870, aged seventy-three. It was written in the *Inspirations-Historie* that "she was a dear, faithful and god-Fearing little mother, patient and composed and trusting God's will with all her many difficulties and sorrows which she in her life had to endure and bear with her disobedient children some of them becoming victims of the world."[11]

THE YEARS PASSED AND MEMORY OF JOSEPH AND HIS SON GOTTLIEB FADED, but examples of their prints and paintings remained in village parlors where change came slowly. In one family attic, children amused themselves on rainy days by drawing on the backs of leftover prints, and also by cutting out the fruits and flowers to paste in their scrapbooks. Sometimes older folk told about the two artists who had made the pictures and who loved plants.

JOSEPH PRESTELE HAS BEEN WELL-DESCRIBED BY CLAUS NISSEN, THE GERMAN art historian, as an outstanding artist in a technically difficult field to which few artists have been willing to devote their lives. In him the skills of an artist were joined with a passionate love of nature.[12]

Prestele's life was a continuous struggle against poverty and ill health, combined with a religious dedication which must have influenced his art but for which much of his artistic career was sacrificed. Two of his sons left their home and their religious community to "enter the world" and he never ceased grieving for them. His last years were spent in obscurity, far from those who appreciated his work, and who could have aided and encouraged him.

As an artist, Prestele had exceptional skills. He could sketch, paint in oil and watercolors, and had the rare ability to engrave on stone. As a colorist he was remarkable.

In Europe, when his career was developing and after his work had appeared in a number of outstanding botanical works, his religious affiliation brought him to America. Here opportunities were limited. Chance—and need—often resulted in his engraving the drawings of others, making illustrations for magazines, or creating the speciality of fruit and flower prints for the use of nurserymen. To all of these tasks, whether artistically interesting or mundane, he brought the same patient care. His opportunities in America may have been limited, but he was one of the small band who lent distinction to the beginnings of botanical art in America.

PART II

The Three Prestele Sons

Joseph Prestele trained his three sons—Joseph Karl Martin (in America referred to as Joseph, Jr.), Gottlieb, and Wilhelm Heinrich (who used the name of William Henry)—as artists and lithographers of plants, but their lives were a disheartening struggle to make a livelihood from a specialty being outmoded by the new methods of color printing introduced during the last half of the nineteenth century. And too, the kind of opportunities which the Asa Grays, John Torreys, and other pioneer botanists had given their father, and the intellectual excitement and encouragement that such men stimulated also faded after the Civil War. Change came to make the skills of the Prestele sons old-fashioned and almost obsolete and they did not, or perhaps could not, adapt.

Wood engraving of Amana, Iowa, about 1870.

4 Joseph Prestele, Jr.

THE ELDEST SON, JOSEPH, JR., WAS BORN IN BAVARIA ON JANUARY 10, 1824.[1] As a young man of twenty-one he came to Ebenezer with his parents in the autumn of 1843 but he soon left the Community to work in Buffalo for a German sausage maker. A few months later he suddenly disappeared, leaving most of his clothes, and telling no one of his whereabouts. Finally he wrote from New York that he was working for a lithographer who was producing a nature book and that he was coloring the plates and earning $3 a week. The father, in writing of this to his parents in Germany in September 1844, bitterly added: "He is an insolent, unruly person yet he is industrious and thrifty when the need arises, and the American freedom suits him well . . . he has several acquaintances in New York."[2]

Joseph, Jr.'s sudden departure for New York may have involved his marriage to one Margaret (born about 1820), who was also a native of Bavaria and a member of the Ebenezer community. Their child Elisabeth was born in New York City, in 1850, and is said to have died there, unmarried, in the 1920s.[3]

Although Joseph, Sr., was bitter about his son's behavior, he remained concerned about his welfare. On July 26, 1850, while the father was working on plates for the Smithsonian's tree book, he wrote Isaac Sprague that "since I am not able to color all the plates by myself and children, I sent a part to New York to my eldest son who lives there and paid him already $32 for coloring plates."

After that, Joseph, Jr., disappears from the record until about 1870 when he and his family came to Bloomington, Illinois, where he joined his brother William Henry in working as a fruit and flower painter for the nurseryman Franklin Kelsey Phoenix. The family lived outside the city limits in a semirural German-Catholic enclave whose residents had large gardens and could keep farm animals, while the men worked in town.[4] The job with Phoenix ended after a year or so, and soon afterward Joseph, Jr., left Bloomington to continue the unhappy course of his life. Neither he nor his family is heard of again.

5 Gottlieb Prestele

THE SECOND SON, GOTTLIEB, WAS BORN IN MUNICH, AUGUST 29, 1827. HE was sixteen years old when he came with his parents to Ebenezer. In the autumn of 1844, when his father took up drawing and painting again, Gottlieb became his devoted pupil and assistant. He won his father's praise for diligence and for having, as Prestele wrote Torrey on April 7, 1845, "much joy and talent for my field of work." He added that Gottlieb was being particularly helpful with drawings and, as examples of his son's work, the proud father sent Torrey six small compositions of garden and wild flowers, and six others of roses, which he offered for sale.[1]

When work began on the tree book for the Smithsonian, Gottlieb was kept busy with coloring and other tasks. But in the autumn of 1851 the work slacked off and the father was concerned that his son would have no employment during the winter. He wrote Gray in the hope that something could be found to keep the young man busy, but nothing may have come of it.[2]

As Joseph's production of nurserymen's plates increased, Gottlieb shared the work with him. Their division of labor is unclear, but occasionally Gottlieb produced a complete plate himself, for a number he made while at Ebenezer have his signature (Chapter 2, under *Plates for Nurserymen*). Later, when he went with his parents to Amana, Iowa, some of the first work that he and his father did there included a number of plates "By J. & G. Prestele, Amana, Iowa" and "Amana Society by J. & G. Prestele. Amana Iowa County Iowa." Gottlieb's individual signature occasionally appears on nurserymen's plates, as with "Drawn from nature & colored by G. Prestele." The son may also have done more and more of the coloring, and also the printing, since, as we have seen, his father disclaimed any practical experience in such work. Also, we know that Gottlieb did print plates after his father's death.

None of the Prestele boys appear to have learned the lithographic process

FIG. 35 Arrangement of Fruit and Flowers. Unsigned lithograph, attributed to Gottlieb Prestele; engraved on stone (with shading in chalk?); 26.5 × 34 cm. (framed). Courtesy, Mrs. Lillian Haldy and Lanny Haldy, Amana, Iowa.

This ornamental print with its soft shading belongs to the small group of decorative plates of fruits and flowers that Joseph and Gottlieb Prestele made and handcolored for framing. None of the examples found have captions or signatures of the artist. The design shown here includes elements from nurserymen's plates, such as the Bleeding Heart, the roses, and many of the fruits. The vase appears to be the same as the one used by Joseph Prestele in his watercolor composition "Blue Vase with Strawberries" (Plate 86).

of engraving on stone, at which their father was such a master. Instead, Gottlieb developed considerable skill in drawing in chalk on stone and, one suspects, in drawing in lithographic ink (tusche) on stone or on transfers (see Page 46; and Chapter 2, note 28). Plates that can be attributed to him were produced in both techniques. His lithographs of tulips, gladioli, and other flowers were drawn in a freer and more simplified style than the meticulously detailed and crisp one used by his father. Some of Gottlieb's watercolor flower paintings are also done in the same freer style. One example is his watercolor painting of an elegant "Cypripedium spectabile. Lady's Slipper (*Frauenschuh*)." (See Plate 88.) Perhaps he too shared his father's dream of painting the wild flowers of America.

When one considers Gottlieb's lifelong isolation from art and from artists

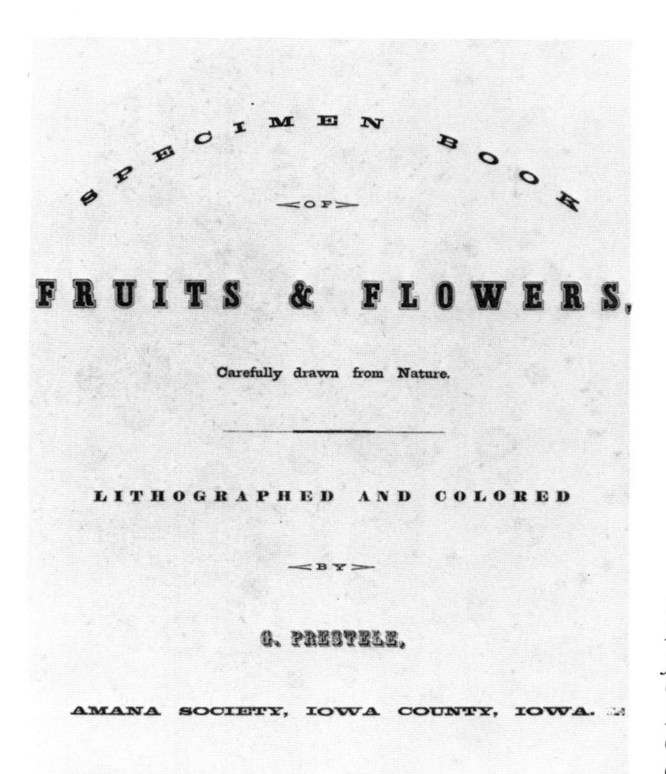

FIG. 36 Undated title page, *Specimen Book of Fruits & Flowers, Carefully Drawn from Nature. Lithographed and Colored by G. Prestele, Amana Society, Iowa County, Iowa.* Courtesy, Special Collections, Ohio State University Libraries, Columbus, Ohio.

other than his father—first at Ebenezer and later at Amana—and the authoritarian restrictions under which he lived and worked, his artistic skill was remarkable.

In his nurserymen's plates Gottlieb, like his father, sometimes used the same design for different subjects which had almost identical shapes, such as fruit. Changes in coloring and sometimes minor variations in details provided sufficient disguise. Surely no deception was intended; certain pears, apples, plums, and other fruits do look very much alike, and the father and son must have been hard put at times to contrive new compositions for illustrating the many varieties of fruit, and particularly different ways of arranging twigs and leaves.

Some years after Joseph, Sr.'s death in 1867, the Amana Society printed an undated title page—*Specimen Book of Fruits & Flowers, Carefully Drawn from Nature. Lithographed and Colored by G. Prestele, Amana Society, Iowa County, Iowa.*[3] Apparently this sheet was intended for binding with assortments of his plates, a service which the Society had begun to offer. A bound volume with this title page now at Ohio State University has a "List of Plates"—handwritten presumably by Gottlieb—for the assortment of twenty-five plates of fruit which the original purchaser had ordered. None of these lithographs carry Gottlieb Prestele's name; many have the same Amana Society imprint. Clues which aid in identifying Gott-

List of Plates.

Red Astrachan - - - - Apple
Fulton - - - - "
Jonathan - - - - "
Willow Twig - - - - "
Fameuse - - - - "
Rawles Jannet - - - - "
Benoni - - - - "
Swaar - - - - "
Wine Sap - - - - "
Domine - - - - "
Baily Sweet - - - - "
Fall Wine - - - - "
Yellow Belleflower - - - - "
Ben Davis - - - - "
Cranberry - - - - "
Large Red Crab - - - - "
Hyslop - - - - "
Bartlett - - - - Pear
Flemish Beauty - - - - "
Cole's - - - - Quince
Clinton - - - - Grape
Concord - - - - "
Wilsons Albany Strawberry.
Green's Prolific - - - - "
Currants 3 different kinds.

FIG. 37 "*List of Plates*." The purchaser of a selection of plates could, if he wished, have them bound at a small additional charge. Often his name, or that of his firm, was stamped on the cover. Courtesy, Special Collections, Ohio State University Libraries, Columbus, Ohio.

lieb's work include the captions on some of the plates which have inept hand lettering, obvious changes, and irregular spacing between words.

Gottlieb's sales were modest, if his notations on several file copies of his plates are any indication. Sixty prints of "The Jefferson Plum" were made on January 11, 1870, and 150 more on December 26, 1875. The "Orange Quince" plate has six printings from April 2, 1863, to January 11 and 12, 1871, totaling 405 plates. A third plate, with the penciled caption "Magnolia," without a varietal name, had six printings from December 2, 1864, to June 28, 1872, resulting in 375 plates. The final note on this last plate, dated February 26, 1873, recorded that the magnolia design had been ground off the stone so that it could be reused.[4]

Sometime in the early 1870s appeared a printed one-page *Catalogue of Fruit- and Flower-Plates. Drawn from Nature. Lithographed and Colored by Amana Society, Homestead P. O., Iowa.* (See Illus., p. 331.) This sheet lists 237 plates of fruits, berries, and flowers. Many of these subjects were on Joseph Prestele's earlier list, and those plates probably represent his and probably Gottlieb's too. The price of rose and grape plates was reduced to 30 cents; the fruits and flowers remained at 25 cents. Bound assortments of 50 and 100 plates were available on order.[5]

The *Catalogue* was further revised about 1875 (Illus., p. 331) by adding new titles and dropping others, resulting in a total of 244 plates then being offered for sale. The price of the fruit and flower plates was set at 25 cents; 15 cents for orders of 50 and more. There was a charge of $1 for binding 50 plates; $2 for 100 or more.[6] Nowhere on these two *Catalogues* appeared the name of the artists.

Sometime after December 1875, Gottlieb's last-known date of production, his sales became so poor that he stopped making nurserymen's plates. He was then assigned to work in the Amana calico printing shop,[7] but he continued painting for his own pleasure, using his homemade paint box with its porcelain mixing cups, its small cans and tubes of color, and its brushes.[8] Like his father, he too gave watercolor paintings as gifts. A watercolor (Plate 87) from this period, mentioned earlier, was a guide to meditation on the omnipotence of God, and an expression of his deeply felt religious beliefs.

Gottlieb's later years may have been rather lonely, for his wife Wilhelmine Zscherny whom he had married in 1868 had died in 1877; there were no children. He seems to have been a quiet, patient, and uncomplaining man who "from his youth, always tried to do good," as was written in the Community's *Inspirations-Historie*. There is no photograph of him. Perhaps none was ever taken—just as none was ever taken of his father or others in his family—because the Community disapproved of such worldly things.

When Gottlieb died on February 6, 1892, it can be assumed that the arrangements for his funeral followed long established custom. After a service in the church the mourners lingered briefly near the church door and then formed a procession to walk to the cemetery at the west side of town. Meanwhile, his coffin, which had remained at his home, was placed in a buckboard wagon used only to convey the dead to their final resting places, and driven slowly away, followed by his relatives. At the cemetery Gottlieb, like others before him and those who would follow, was laid to rest in a long row of graves where the dead are placed in succession as their time comes. Their heads are to the west; they face east in preparation for resurrection day. Each grave is marked with a modest stone of identical size. Over the years the stones have sunk lower and lower into the turf. As evening comes, their lengthening shadows join those of the pines and spruces surrounding that quiet place.

6 William Henry Prestele

THE THIRD AND YOUNGEST SON, WILHELM HEINRICH, OR WILLIAM HENRY AS he was known in America, was born in Hessen-Darmstadt on October 13, 1838.[1] When he was five, he came with his parents to America and spent his youth at Ebenezer, New York. He was a spirited youngster who caused much talk. Once, when he was about fourteen, he was one of the young people admonished by Christian Metz to "Remove from your heart all mischief, all abominations and bad habits, and draw nearer to the Lord," the only one in the group specifically named.[2] But, as a present member of the Community recently commented, "They admonished everyone!"

Some years later, but before 1858 when his family moved to Iowa, William Henry left Ebenezer to "live in the world," which for him meant New York City, where his brother Joseph had gone. What work he found during his first years there has not been learned but shortly after the outbreak of the Civil War in 1861 he joined Company E, 26th Regiment, New York Volunteers, and from time to time wrote his father. A few pages from these letters, and several envelopes have survived. One letter is dated 1862; an envelope locates his company at Alexandria, Virginia.[3] When his enlistment expired, he returned to New York. He married then (or earlier?) an Irish girl, Ann, born in 1843. Their daughter Margaret was born in 1864.[4] The New York city directory for that year lists him as an artist living on Ninth Avenue. About 1867 the course of his life was suddenly changed when he was hired to make a line of nurserymen's plates by Franklin Kelsey Phoenix, owner of the Bloomington [Illinois] Nursery.

"F. K. Phoenix," as he was generally known, was an interesting mix of the practical and the visionary. By nature he was firm, honest, and kindly. "He loved everything that pertained to the elevation of the human race," he was quoted as saying by a writer for the Bloomington *Weekly Pantagraph*, who added that Phoenix was an open and avowed enemy to strong drink of every kind, including tea and coffee, nor did he use tobacco.[5]

His family had moved from New England to western New York State, and then moved again to be among the pioneers of Wisconsin, settling at Delavan in 1837. Franklin started a nursery there about 1842, when he was seventeen, and operated it with apparent success. Ten years later he bought a small nursery in Bloomington, Illinois, where he moved in 1854 after winding up his Wisconsin affairs. Phoenix stressed "Western trees for Western orchards."[6] He collected and tested fruits, berries, and ornamental plant material, seeking those that could survive the harsh climate of the upper Midwest, even personally selecting promising stock in Europe. His advertisements and catalogues had a wide circulation, including the South and the Far West. He was frequently quoted in the horticultural and agricultural press. A biographer claimed that his "eloquent appeals for trees for profit, trees for pleasure, trees for future generations," inspired pioneers settling on the treeless prairies of the West to plant "orchards, groves and windbreaks all over the western states."[7]

Phoenix first tried to sell directly to customers, hoping to eliminate the use of traveling salesmen, whom other nurseries had found both a source of profit and of problems, but his losses taught him how effective such agents could be. As settlement of the western states and territories increased after the Civil War, success seemed to reward all his efforts. His nursery grew from the original "five-acre patch" to 600 acres, with large greenhouses and office buildings. Some 500 men were employed during the busy season, and his annual sales grew to $200,000.

When Prestele joined the nursery, it was on its way to becoming the second largest in the United States.[8] In seeking new ventures, Phoenix decided to introduce his own line of nurserymen's plates, an idea probably inspired by the success of D. M. Dewey. Few details survive about how he developed his plans, but important to them was his decision to employ the Prestele brothers—first William Henry and later Joseph, Jr.—to assist in the project.

Unfortunately, we don't know what arrangements Phoenix made with William Henry, who moved to Bloomington with his family about 1867[9] and our information about Phoenix's operation is only fragmentary. It appears that Phoenix was cautious in planning his new venture, and avoided making a major investment in lithographic equipment by having his plates printed by firms in the Midwest. Prestele superintended the limited operations at the nursery, chiefly the preparation of designs for the plates, and their coloring after they were printed.

No plates have been located which we are certain were made under Prestele's supervision. The only examples of plates issued by Phoenix which are known to this writer are in an assortment bound in 1876 or later, with *Bloomington Nursery* on the cover. Of these thirty-eight plates, sixteen have the imprint "F. K. Phoenix, Bloomington, McLean Co., Ils." and the names of the lithographing firms in

FIG. 38 "*Lonicera.*" "I. Percyclimenum. Woodbine." (*L. periclymenum.* Woodbine Honeysuckle. Woodbine.) "II. Sempervirens. Monthly Honeysuckle." (*L. sempervirens superba.* Scarlet Trumpet. Honeysuckle; Coral Honeysuckle.) "III. *Frazerii.* Strawcol[ore]d Honeys[uck]le." (*L. sempervirens sulphurea.* Yellow Trumpet Honeysuckle.)

Unsigned lithograph, attributed to William Henry Prestele, Bloomington, Illinois, ca. 1870; drawn in "tusche" (lithographer's ink) on stone (?); colored; 27.5 × 21 cm.

This print is included in a nurserymen's plate book with a cover label of the Bloomington Nursery, Bloomington, Illinois. William Henry Prestele was employed by F. K. Phoenix, proprietor of the Bloomington Nursery, to direct the production of nurserymen's plates from about 1867 to 1871 or 1872. Attributions of plates to him can only be made with uncertainty; the quality of the Phoenix plates is not impressive. Author's collection.

Chicago, Louisville, St. Louis, and Milwaukee that printed them. These can be dated from about 1872 until about 1876. Another fifteen lack the names of their printers, although they were the work of outside firms. Seven have the Phoenix imprint. Most, if not all, of these plates date from after Prestele left the nursery.[10]

When the plates in this small assortment are examined under high magnification, it becomes apparent that several different techniques were used in their production. Some designs are drawn and shaded in "chalk" on stone. Others were drawn in "tusche," the lithographer's ink, either directly on stone or, as seems more likely, on transfer paper used in lithographic work, from which the image could be applied directly to a stone for printing. Thus drawings could have been sent to the printers for copying in chalk on stones, or finished drawings in "tusche."[11]

Prestele's initial efforts would have been directed toward building up a stock of plates. Shortly after the project was launched it was reported that the nursery had a stock of about a thousand subjects on hand which, if true, was an astonishing number after so short a preparation time.[12] Certainly, William Henry couldn't have drawn all of these from nature if, indeed, he drew any of them in that way. Instead, doubtlessly under pressure from Phoenix, he took shortcuts. At least six plates in the bound assortment were copied from his brother Gottlieb's plates with only minor changes. Possibly Phoenix augmented his initial stock by buying finished plates from Gottlieb.

In the spring of 1869, Phoenix launched his new project. The following August, Thomas Meehan, the publisher and editor of the *Gardener's Monthly and Horticultural Advertiser* (Philadelphia) commented on the Bloomington Nursery's new plates. "We have now before us a fruit piece . . . prepared by W. H. Prestele. We are in the habit of admiring European art in this line, and have often wished Americans could successfully compete with it. *We now have it here*. We never saw anything of the kind better executed from any part of the world. We wish the new enterprise every success."[13] This was not hollow praise; Meehan was not given to such. He knew and appreciated good art as was demonstrated by the beautiful color plates he was then publishing in his horticultural magazine. It is inconceivable that the plate or plates that Meehan saw were like those in the bound collection. Possibly he saw a plate made by William Henry's father, or by his brother Gottlieb.

Sales of the Phoenix plates were reported to be brisk by a Bloomington reporter in March 1870.[14] His tour of the nursery included the "drawing and coloring department" adjoining the main office, which was "under the supervision of Wm. H. Prestele, of N.Y., an artist of acknowledged talent and a most pleasant gentleman." Unfortunately, the reporter's account is rather vague because his attention was distracted by "four or five bright, rosy-cheeked girls busily engaged in drawing and coloring" who also "found time to take an occasional peep from the corners of roguish looking eyes." He did manage to comment on the "beautiful, sweet, and so perfect" plates.

The reporter also mentioned that Phoenix planned in the autumn to double the number of artists employed. Those plans may have included adding Joseph Prestele, Jr., to the staff, for he came to Bloomington about 1870, presumably bringing his wife and daughter with him.[15]

In 1871 Phoenix advertised "Colored lithographs of Fruits and Flowers, superior to all others. Four samples by mail for $1. Send for list."[16] The size of the plates is not given, but apparently they were large ones, priced competitively with Dewey's. In 1875 Phoenix was selling his for 25¢ each; bound collections of 100 were $21.40. In 1878 Dewey charged from 25¢ to 60¢ for his large prints, 11 by 9 inches (28 by 23 cm.), subject to special discounts.[17]

The Prestele brothers ended their association with Phoenix late in 1871 or early in 1872. We don't know what caused the break. However, the production of plates at the nursery appears to have continued under the direction of Albert H.

Foote, who had started at the nursery as a laborer in 1870 and by 1873 had become an "Artist." Three years later, his luck changed and he lost his job.[18]

The small number of Phoenix Nursery plates that we have seen are disappointing.[19] They suffer from hasty draftsmanship, careless tinting, and unnatural coloring. Some have crudely lettered captions as though they were proofs and not finished plates. We can credit most of these to Foote's regime at the nursery, but lacking plates which we know were made under Prestele's supervision, we have no way of judging the quality of his output. However, the plates William Henry made and signed some years later show his limitations as an artist, a shortcoming that he overcame in time.

After the rupture with Phoenix, William Henry stayed on in Bloomington, going into business with an L. B. Littlefield, publishing fruit and flower plates. Since Prestele had no lithographic press, it is possible that Littlefield had one. The venture, however, was shortlived.[20] Next, Prestele tried to carry on by himself but how he did so without a press is not explained. That too soon ended. Meanwhile, his brother Joseph left Bloomington for parts unknown, perhaps to return to New York.

While William was self-employed in Bloomington, he produced a mammoth "Showcard," which he had printed in St. Louis in 1872. The design is that of a richly colored wreath of fruits and flowers, a lavish portrayal of nature's bounty. The oval center was left partially blank so that the name of a nursery, or a merchant, could be inserted. Twelve years later, moved by some special act of kindness shown him by Charles Downing of Newburgh, New York—the horticulturist, author, and brother of Andrew Jackson Downing—Prestele wrote in the center of one of these showcards his expression of gratitude for Downing's aid during his "hours of severest trials," and sent it to his benefactor.[21] The inscription reads:

> *To Charles Downing*
>
> *With kind feeling and a heart full of everlasting gratitude, this collection of American Fruits and Flowers is presented to you, as a token of esteem, by your humble grateful friend, by whom your noble and true friendship, during the hours of his severest trials will always be remembered and cherished with a warm heart, as long as life will remain, and may God our heavenly Father pour down his blessings hundredfold on you, and grant you many years yet to come with bountiful peace and happiness. This will always be the prayer of your humble and ever grateful friend, W. H. Prestele.*
>
> *Charles Downing Esquire from his Friend W. H. Prestele,*
> *Iowa City, Ia. April 26th., 1884.*

Little information has survived about William Henry's wife and children

FIG. 39 Hand-painted Card in a Gift Portfolio of Nurserymen's Plates.

In 1876 William Henry Prestele, ever sentimental and warmhearted, assembled a portfolio which he inscribed: "Presented to Zach Seaman by his friend Wm. H. Prestele. 1876." Ten years later, perhaps as a wedding or anniversary gift, William Henry painted the decorative card (8.5 × 10 cm.) shown here, which was added to the portfolio.

This portfolio of nurserymen's plates originally contained 130 examples, of which 39 remain. Of these, 1 has the signature of the Amana Society, Homestead, Iowa; 8 have William Henry Prestele's Iowa City signature; and the remaining are unsigned but can be attributed to either Joseph Prestele, Sr., or Gottlieb Prestele. The plates, 29.5 × 23 cm., are mounted between protective sheets of blue tissue in a book that was bound by the Republican Press, Iowa City, about 1876. Courtesy, Mrs. Emma Setzer, South Amana, Iowa.

during their Bloomington years. We only know that two children were born there, Francis J. (around 1869–70) and Emma C. (around 1871–72) and that his wife Ann died sometime later, either in Bloomington or after the family moved to Coralville, Iowa, about 1875.[22]

Early in 1870 Prestele went to Amana to visit his widowed mother, who had been weak and ill for several years and whom he had not seen since he had left Ebenezer more than ten years before. On January 6, an hour after he had left her to return to Bloomington, she died.[23]

About 1875, William Henry moved his family from Bloomington to Coralville, a village near Iowa City and not far from the Amana community. There he brought his second wife, Susanna Gefäller of Homestead (one of the Amana villages), whom he married about that time. Their daughter, Lilly, was born some three years later.[24]

William Henry set up as a lithographer in Iowa City, having acquired a lithographic press, stones, and all the other paraphernalia used in his trade. How he was able to make such a sizeable investment is not clear, but it is more than

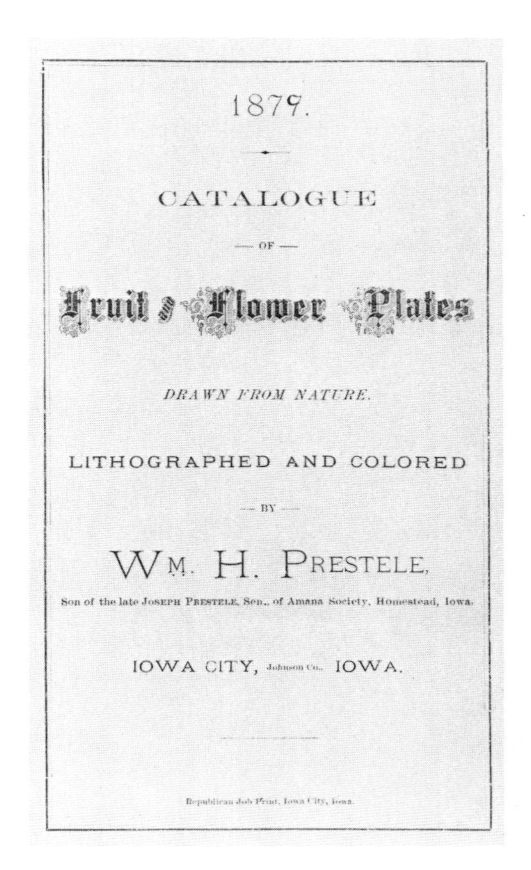
1879.

CATALOGUE

— OF —

Fruit & Flower Plates

DRAWN FROM NATURE.

LITHOGRAPHED AND COLORED

— BY —

WM. H. PRESTELE,

Son of the late JOSEPH PRESTELE, Sen., of Amana Society, Homestead, Iowa.

IOWA CITY, Johnson Co. IOWA.

Republican Job Print, Iowa City, Iowa.

FIG. 40 "187[9]. *Catalogue of Fruit & Flower Plates Drawn from Nature. Lithographed and Colored by Wm. H. Prestele, Son of the late Joseph Prestele, Sen., of Amana Society, Homestead, Iowa. Iowa City, Johnson Co., Iowa.*" Republican Job Print, Iowa City, Iowa. Author's collection.

possible that the equipment came from his brother Gottlieb, who retired from the plate business about that time. In addition, William Henry seems to have gotten a sizeable stock of finished plates from his brother to supplement his own production.

To launch his plate business William Henry published a catalogue listing the 472 titles of plates he had in stock. The large number reinforces the belief that many of them came from Gottlieb. The catalogue was printed with the date "187 ", the final digit being left blank so he could fill in the year and thus continue using the same edition.[25] The copy which this writer has seen is dated 1879 and contains twenty-two additional titles of plates, added to the catalogue in Prestele's small, neat handwriting. This catalogue seems to have been his only effort to promote his business; no advertisements have been found in Iowa gazetteers or Iowa City directories.[26] In 1880 his eldest daughter, Margaret—now called Elisabeth—was assisting him as an "artist's apprentice."[27]

In the introduction to his catalogue, William Henry stated with obvious pride that he was the only successor to his father's business and the only person "that can furnish the genuine *Prestele Lithographed Colored Plates.*" Since his removal to Iowa City he had assembled "a good stock of Plates, Lithographic Printing Press, Stones, &c.," and had added many new and popular varieties of plates and improved old ones.

William Henry advertised plates in four sizes: large (12 by 9 inches), selling for 15 and 20 cents; a pocket edition (9 by 6 inches), at 15 cents; and two small editions (8 by 5, and 7 by 5 inches)—the reason for making editions so near in size was not explained. Bound collections of forty plates sold for $6.50 (large size), $6 (pocket edition), and $5.50 and $5 for the small editions. The plates in stock are listed individually in the catalogue. He announced that he made "original drawings" from nature free for his regular customers; others were charged 50 cents for the first copy and the regular catalogue price for additional copies. Examples of his plates, owned in the Iowa City area, have the signatures "By Wm. H. Prestele, Artist, Iowa City, Iowa"; "Drawn, Lith. & Col^d. by W. H. Prestele, Iowa City, Iowa"; and most frequently "Lith. & Col^d. by W. G. Prestele, Iowa City, Iowa."

William Henry was very proud of his "New Show Card" which, from his description, was even more elaborate than his first one, described earlier. He announced that the design included "a Nursery, Greenhouses, Dwelling House, Lake, Mountains, etc., seen from a verandah, with Table in centre, covered with choice Fruit, on each side the Queen of the Prairie Rose, twining up on a lattice work, in full bloom; and, prominent of [over?] all, the *American Eagle* on a full stretch, up in the air, carrying a large, broad ribbon in his bill, with advertisement on it, FRUIT AND ORNAMENTAL TREES, etc., in large, full type. A superb picture, brilliantly colored by hand, size 19 × 24 inches. Price, $5.00 each, $33.00 per dozen."[28]

We don't know how many of these showcards he sold, or how well he did with his nurserymen's plates, but between those sales and his orders for billheads, stationery, and other commerical work, he managed to carry on his business in Johnson County, Iowa, for more than a decade.[29] His second wife, Susanna, died at their home in the suburb of Coralville, aged forty-three, on March 27, 1882.[30]

Then, six years later, on August 1, 1887, when he was forty-nine, William Henry's life was dramatically changed again, this time by his appointment as the first artist on the staff of the newly created Pomological Division of the Department of Agriculture in Washington, D.C. Fruit growing had become a major crop in America, and the division was intended to serve the interests of the growers by a broad program, including research in those fruits which could be successfully grown here. Accurate records, including botanically detailed watercolors of the fruits, were an important element in carrying on such work.

Prestele's beginning salary was $1,000 a year, one-third of the budget of the entire division which then had a staff of three. By 1894 he was receiving

$1,200.[31] Within two years of his appointment, five more artists were added.

Prestele's assignment was to make drawings, some in India ink and some in watercolor, "of both the exterior and interior of the fruit, with the leaves and twigs characteristic of each." "These," the research-minded chief of the division, G. Onderdonk, commented in his *Report* for 1887, "are invaluable for comparison and reference, and a portion for publication."[32] Each of the pomologist's reports for the years 1887 to 1892 include from six to eight chromolithographs of watercolors signed "Wm. H. Prestele fecit." In 1893 only one plate by him appeared; the rest are by another hand. Various firms did the printing, including A. Hoen and Co., Baltimore.[33]

One of Prestele's special assignments in the division was to make carefully researched watercolors of some thirty species and subspecies of native grapes intended as illustrations for a monograph by T. V. Munson of Denison, Texas, a leading authority on native grapes, who was commissioned a special agent in the division of pomology for that purpose. The project took six years to complete, requiring extensive travel by Munson to collect vines from the wild, and painstaking work by Prestele in preparing drawings from dried specimens. Sometimes Prestele found the available specimens inadequate and, as in a hastily penciled interoffice note to his division head, he asked for better samples.

The artist's preliminary color sketches were reviewed by Professor Munson who, when necessary, suggested corrections. The final watercolors are about 14 by 17 inches, beautifully designed and drawn, with the natural colors carefully reproduced in all their subtleties, making plates of exceptional beauty and scientific value.

Munson's text and twenty-nine watercolors by Prestele were completed in 1891 but the monograph was never published because of high production estimates. Since then most of the Munson text has disappeared, but Prestele's original watercolors are preserved at the United States Department of Agriculture's Research Center, Beltsville, Maryland.[34]

On May 7, 1894, when William Henry was fifty-six, he completed a watercolor of the "Hartford Prolific" grape which he signed and dated; thereafter he disappears from the record. There is a story in Amana that he had a stroke.

Any appraisal of William Henry's work must be limited because his work was so inconsistent in quality, and so many artistic and technical questions about it remain unanswered. As mentioned earlier, many of his plates from the 1870s were very ineptly done. Even his best examples from that period lack the brilliance of his father's finely engraved plates, partly because William Henry worked in a less delicate technique, that of chalk on stone, the usual lithographic process. Then too, he was not always as patient or as adept in coloring as his father and his brother Gottlieb. But if they were more painstaking, and more technically skilled, they nevertheless lacked William Henry's creativity. Their nurserymen's plates followed a similar pattern; a specimen of fruit, or blossom, with stem and leaves. William Henry varied his compositions; a climbing rose is shown on a trellis; his showcards are extravaganzas. He apparently signed all of his work—and in bold letters. Yet he could be subtle too. The watercolors he made near the end of his life for the United States Department of Agriculture demonstrate the ability he developed over the years as an artist. Their compositions are natural and graceful, and the subjects are carefully portrayed with an attention to botanical details that would have delighted his parent. Reproductions of many in the department's annual *Reports* are handsome. William Henry's art, like the man himself, is not easy to categorize. From our infrequent glimpses of him, we sense that he was a free spirit about whose life and work we would like to learn more. To the end he remained as his father had described him as a youngster, "Skilled and clever, lively and active."

Notes

Preface

1. Dr. Fritz H. Herrmann to the author, October 25, 1982. The articles by Drs. Fritz H. and Lore Herrmann are cited in Chapter 1, note 3; and Chapter 2, note 22. On September 2, 1845, Joseph Prestele wrote Asa Gray that he was sending him a copy of his new work on the most poisonous plants of Germany (*Die wichtigsten Giftpflanzen Deutschlands*). Presumably this is the copy preserved at the Gray Herbarium of Harvard University—the only copy which this writer has located in America. The work was published in two parts, each with the same text on the printed covers. The portfolio (43 × 27 cm.) contains Prestele's engravings; the smaller publication (19.7 × 13 cm.) has Soldan's text consisting of a Foreword (four pages), Table of Contents (one page), and a descriptive catalogue of the individual plates (fifty-six pages). A fragmentary copy of the portfolio is in the collection of the Amana Heritage Society Museum of Amana History, Amana, Iowa. For a discussion of the work see Chapter 1, p. 33.

Chapter I

1. Both Dr. Fritz H. Herrmann (see the Preface; and Chap. 1, n.3), and Thieme-Becker, *Allegemeines Lexicon der bildenden Künstler* (Leipzig: E. A. Seemann, 1908–50), give November 8, 1796, as the date of Prestele's birth, but the register of baptisms at St. Martin's Church, Jettingen, notes that Prestele was *baptized* on that date. The *Inspirations-Historie*, the Amana Heritage Society, Amana, Iowa, (see Chap. 1, n. 10) gives Prestele's age at his death on March 9, 1867, as seventy years and four months, which would place his birth in November 1796. Prestele's discharge from the Bavarian army in 1823 states that he was born in 1796, and gives his name as "Martin Prestele." (Photocopy of Prestele's baptismal record kindly supplied by Dekan Horst Grimm, pastor of St. Martin's Church, Jettingen, Bavaria, West Germany, November 1983.)

2. In the manuscript collection of forty-eight watercolors entitled *Icones plantarum selectarum horti regii Academici Monacensis vivis coloribus pictae a Fr. Jos. Prestele* which Prestele produced in 1812, he signed his name Fr[anz] Jos[eph] Prestele. Claus Nissen, *Die botanische Buchillustration* (Stuttgart: Hiersemann Verlag, 1951), vol. 1, p. 264, states that Prestele's full name was Franz Joseph Ulrich Prestele.

3. My description of Jettingen is based upon data very kindly provided by Frau Hermine Stickel, Stuttgart, West Germany, and Herr Studienrat Theodor Schnürle, living near Ulm, West Germany. Included in the material they sent me was a photocopy of an early nineteenth-century lithographic view of Jettingen ca. 1830, but little changed from when Joseph Prestele lived there. Wishing to reproduce the print in my book and needing a clear photograph of it, my search led to Dekan Horst Grimm, pastor of St. Martin's Roman Catholic Church, Jettingen, who very generously supplied my need. He also sent a photocopy of the Prestele baptismal entry in the church register, and other data—kindnesses which I greatly appreciate. The print, about 44 × 60 cm. with margins, is titled "Markt Jettingen, Sitz des Freiherrsch von Stauffenbergschen Patromonialgerichts 1, Klasse" (Market Town of Jettingen, seat of Baron von Stauffenberg's hereditary jurisdiction). It was drawn from nature by Nikolaus Lechner, and drawn on stone by Gustav Kraus; ca. 1800.

The account of Prestele's early years, given here and following, is based upon Fritz H. Herrmann's carefully researched "Zur Lebensgeschichte des Pflanzenmalers Joseph Prestele," *Sonderdruck aus Wetterauer Geschichtsblätter* (Friedberg/Hessen, 1979) vol. 28.

4. *Cyclopedia of Painters and Paintings*, edited by John D. Champlin, Jr. (New York: Charles Scribner's Sons, 1892), vol. 11, p. 392.

5. New York: Farrar, Strauss and Giroux, 1979, p. 255.

6. As in E. Bénézit, *Dictionnaire critique et documentaire des peintres, sculpteurs, dessinateurs et graveurs*. 8 vols. (Paris: Libraire Grund, 1948–55), vol. 7, p. 19.

7. Joseph Prestele's portfolio is preserved in the manuscript division of the Bavarian State Library, Munich, West Germany. The selections from that work, illustrated here by permission, were obtained with the helpful assistance of Frau Liselotte Renner, Handschriften u. Inkunabelsammlung, of the Library, and Mrs. Richard E. Schroeder, formerly of Washington, D.C., and now a resident of Munich. Dr. Herrmann's comment about the portfolio is in his article, op. cit., n. 3, p. 151.

8. Fritz H. Herrmann, op. cit., n. 3.

9. The discharge paper is owned (1982) by Dr. Louis A. Clemens, Homestead, Iowa, a Prestele descendant.

10. The account of Karolina Russ Prestele is given in the Amana Society's *Inspirations-Historie*, Vol. I for 1867–76 (Amana, Iowa: 1900), pp. 187–88. (Courtesy, The Amana Heritage Society. Copied by Henrietta M. Ruff; translated by Elisabeth Kottenhahn. The following description of the *Inspirations-Historie* was kindly supplied by Lanny Haldy, director of the Amana Heritage Society's Museum of Amana History.)

The *Inspirations-Historie* is the standard source, in German, of the history of the Community of True Inspiration in Germany and America for the years 1714 to 1923. The neatly bound set of ten volumes was published at Amana from 1884 to 1926. The series had two author/editors, the first of whom was Gottlieb Scheuner (1836–1897) who drew on primary sources to write a sensitive and interpretative history of the Community from its beginnings in Germany in 1714 until 1897. Following his death the work was carried on by George Heinemann who continued the chronicle for the years 1898 to 1923 inclusive. His accounts are less literary, more straightforward reports of church services and obituary notices in which most of the biographical information on members of the Community is found. Very few testimonies by the *Werkzeuge* are included in these volumes as the testimonies were printed in separate collections; Scheuner usually included only the important excerpts.

Because of the rather confusing numbering of the volumes, in citing volumes here and following, the dates covered in each volume are given.

The long German title of volume 1 reads, in translation: "Inspiration history or historical report of the founding of prayer meetings and congregations; how they have been commanded by the Lord's Spirit through the Inspiration-Instruments [agents] . . . as well as what happened to the Inspired therein named, within and with the true Inspiration-Congregation during J. F. Rock's lifetime as well as afterwards until the time of the New Awakening in the years 1817 and 1818." Succeeding volumes were all titled *Inspirations-Historie* but with different descriptive subtitles.

11. The children of Joseph and Karolina Russ Prestele were: Joseph Karl Martin (b. January 10, 1824; d. after 1872, place unknown); Karolina Amalia Barbara (b. April 2, 1825; d. 1833); Gottlieb (b. August 29, 1827; d, Amana, Iowa, February 6, 1892); August (d. June 19, 1828); Maria Franziska Elisabeth (b. May 20, 1829; d. July 30, 1833); Franziska Augusta (b. April 15, 1830; d. July 26, 1834); Elisabeth Karolina Franziska (b. August 31, 1831); Caroline [sic] Elise (b. January 20, 1835; d. March 1858); Wilhelm Heinrich [William Henry] (b. Hessen-Darmstadt, October 13, 1838; d. probably in Washington, D.C., about 1894; see Chap. 6). Extracts from the *Inspirations-Historie* supplied by Henrietta M. Ruff, op. cit., n. 10; Fritz H. Herrmann, op. cit., n. 3, p. 140; and the *Register of the United States, containing a list of the officers and employees in the civil, military, and naval service on the first of July, 1895* (Washington: Government Printing Office, 1895) Vol. 1, p. 814. William Henry Prestele's birthdate supplied by Dr. Louis A. Clemens, op cit., n. 9. The death dates of Karolina Amalia Barbara and of Franziska Augusta were included in a watercolor sketch of their graves made by Joseph Prestele (Fig. 6).

12. Fritz H. Herrmann, op. cit., n. 3.

13. Fritz H. Herrmann, "Joseph Prestele."

14. Fritz H. Herrmann, "Joseph Prestele."

15. A history of the Community of True Inspiration, particularly the American portion, is found in Bertha M. H. Shambaugh, *Amana That Was and Amana That Is* (New York: Arno Press, 1976; first published 1932). For a brief history of the Inspirationists and their settlements at Ebenezer, New York, and Amana, Iowa, see: G. Schulz-Behrend, "The Amana Colony," *American-German Review* VII, no. 2 (December 1940): 7ff.

16. Shambaugh, *Amana*, p. 53.

Chapter 2

1. Frank J. Lankes, *The Ebenezer Society*, reprint of 1949 edition (West Seneca, New York; West Seneca Historical Society, 1963), pp. 16ff. In 1941, Lankes, then a resident of West Seneca, located in the former Ebenezer settlement of the Inspirationists, learned about that sect. His growing interest and patient research in their history resulted in *The Ebenezer Community of True Inspiration*, published by the author in 1949, and reprinted by the West Seneca Historical Society, Gardenville, New York (in 1982?). In addition to its informa-

tive text, it contains illustrations and charts of considerable interest. *The Ebenezer Society*, which Lankes also published himself in 1949, appears to be an abridged version of the larger work. Much of my account of the Ebenezer years is summarized from Lankes's two works.

2. See n. 1 and *The Ebenezer Society*, pp. 22–58.

3. Prestele to John Torrey, April 7, 1845. Unless stated otherwise, the Prestele letters to Torrey are in the John Torrey Papers, Library of the New York Botanical Garden, and are quoted with permission.

4. Prestele, Ebenezer, New York, to his father and mother in Germany, October 15, 1843. This first letter to his family, written after his arrival at Ebenezer, described his journey, which I have summarized here. Unless stated otherwise, the quotations in the text regarding this journey are from this letter. Original letter owned by the Amana Heritage Society which generously provided a translation by Mrs. Magdalena Schuerer of Amana, and gave permission to publish excerpts here and following.

5. Lankes, *The Ebenezer Society*, p. 59.

6. Prestele, Ebenezer, New York, to Asa Gray, January (8?), 1846. Unless stated otherwise, the Prestele letters to Gray are in the collection of the Gray Herbarium of Harvard University and excerpts from them are printed here with permission. No letters by Gray to Prestele are in the Gray archive at the Gray Herbarium, and none have been located elsewhere.

7. Lankes, *The Ebenezer Society*, p. 70.

8. Ibid., p. 58.

9. Ibid., p. 35.

10. Ibid., pp. 88–89, 98–99.

11. An observation by Lankes, ibid., p. 39.

12. The introduction of the limited form of communism is discussed by Lankes, ibid., pp. 61–67, and Bertha M. H. Shambaugh, *Amana That Was and Amana That Is* (New York: Arno Press, 1976; first published 1932), see Chap. 1, n. 15, pp. 59–62.

13. Celibacy ("self denial and the death of old Adam") and marriage among the Inspirationists is discussed by Shambaugh, pp. 117–24.

14. Prestele to his father, August 25, 1844.

15. Shambaugh, *Amana*, pp. 59–62. This was only one of a series of divine revelations reported by the leaders as the move toward communism developed and as dissension grew among some of the brethren.

16. Lankes, *The Ebenezer Society*, pp. 67, 98–99.

17. Lankes. *The Ebenezer Community of True Inspiration*, pp. 83ff.

18. Henrietta M. Ruff, Amana, Iowa, to the author, July 14, 1980.

19. Prestele to his father, August 25, 1844.

20. Ibid.

21. Ibid.

22. Prestele to Carl Christian Bindernagel, May 27, 1844, quoted in Lore Herrmann, "Die wichtigsten Giftpflanzen Deutschlands," *Sonderdruck Wetterauer Geschichtsblätter*, vol. 26 (Friedberg/Hessen, 1977), p. 283. Translation by Elisabeth Kottenhahn.

23. Prestele to his father, August 25, 1844.

24. Address on Prestele's letter to Gray, January (date omitted) 1845. Gray noted its receipt February 7, and that he sent a copy of it to Dr. Torrey.

25. Prestele to Gray, January (?), 1845.

26. Gray to Torrey, March 8 [1845]. Reprinted in Jane Loring Gray, ed., *Letters of Asa Gray*, 2 vols. (Boston and London, Macmillan and Co., 1893), 1:330.

27. W. D. Richmond, *The Grammar of Lithography* (London: Wyman & Sons, 1880), p. 131.

28. For descriptions of these processes see: Aloys Senefelder, *The Invention of Lithography*, translated by J. W. Muller (New York: The Fuchs & Lang Manufacturing Co., 1911), pp. 200ff; Michael Twyman, *Lithography 1800–1850* (London: Oxford University Press, 1970), pp. 104, 115ff; and Richmond, *The Grammer of Lithography*, op. cit., n. 27, pp. 3, 7, 131ff.

The dry point or engraving on stone technique is also described in *Nouveau Manuel Complet de l'Imprimeur Lithographie*, by L.-R. Bregeault (new revised edition with additions by Knecht and Desportes, Paris, 1850), pp. 69–70. This technique is quicker and less expensive than engraving on copper or steel, from which technique it was adapted; the lithographic stones are not worn down as quickly by printing and give many good impressions as the metal plates.

The description of engraving on stone given here is greatly condensed from accounts in several of the works cited above and from data kindly provided by Majorie B. Cohn, associate conservator, Fogg Art Museum, Cambridge, Massachusetts, Harvard University.

29. Prestele to Torrey, April 7, 1845.

30. On the first of these plates is written: "Prestele. Specimen of his engraving."

31. G. & W. Endicott, lithographers, New York City, 1845–48. George Endicott (b. Canton, Massachusetts, June 14, 1802) worked as an ornamental painter in Baltimore c. 1828 when he first took up lithography. In April 1830 he joined with Moses Swett to form Endicott & Swett,

which moved to New York in December 1831. Swett left the partnership c. 1834 to form his own business. Endicott continued alone until 1845, when he was joined by his younger brother William Endicott (b. Canton, Massachusetts, August 20, 1816). The firm of G. & W. Endicott lasted until the death of George, April 21, 1848, and then became William Endicott & Co., until the death of William, October 18, 1851. The succeeding firm, Endicott & Co., formed in 1852, continued until 1886.

By the mid-1840s George Endicott, with his "shoulder-length silvery hair," had become the "revered 'old man' of the New York lithographic community," despite the fact that he was only forty years old. The firm began producing chromos between 1845 and 1849 but continued to turn out hand-colored prints into the 1860s. "The Endicotts were unquestionably highly skilled lithographers," according to Peter C. Marzio in *The Democratic Art: Chromolithography 1840–1890* (Boston: David R. Godine, 1979), pp. 42–43; see also George C. Groce and David H. Wallace, eds., *The New-York Historical Society's Dictionary of Artists in America 1564–1860* (New Haven: Yale University Press, 1957).

32. As a youth, Isaac Sprague (b. Hingham, Massachusetts, September 5, 1811) had been apprenticed as a carriage painter with an uncle. His hobby was drawing and painting plants and birds. John James Audubon saw some of his work, liked it, and sometime later invited Sprague to study with him in New York. This led to Sprague's accompanying Audubon on an expedition to the upper Missouri River to assist in making drawings. After 1844, Sprague first did work for Gray; in 1845 he moved to Cambridge and became Gray's principal illustrator. He also worked for many others, including the U.S. government, preparing botanical drawings for reports of various western expeditions. In 1847, Gray persuaded Sprague to make the folio drawings for the botanical report of Captain Charles Wilkes's Pacific expedition (1838–42) for $10 each, including "full dissections" of the plants, the price being "as low as Sprague can do them for, to any advantage, even if he had nothing else to do . . . I can only hope that the experience and facility he is getting will enable him to knock them of faster hereafter." (Gray to Torrey, July 20, 1847, reprinted in Jane Loring Gray, ed., *Letters of Asa Gray*, op cit., n. 26, vol. 1, p. 347.)

Sprague and Gray had a long association, but Gray, always a perfectionist with boundless energy and a mind whirling with ideas, sometimes grew impatient with the artist's deliberate pace and his frequent ailments. In 1849 Gray wrote that Sprague said he could work "but a little while at a time, from a difficulty in breathing . . . that his health and vigor [was] giving way" (ibid., p. 364). In 1853 Gray complained that "Sprague is too slow, and too feeble in health, to do half of what I want done, let alone others. I must import an additional draughtsman." But despite his fuming and his impatience, Gray respected Sprague's "gift for beauty, his accuracy, his quick appreciation of structure, and his skill in making dissections," or so Mrs. Gray wrote after her husband's death (ibid., p. 329). Whatever Sprague's physical problems may have been, he outlived Gray by some seven years, dying March 15, 1895. "This modest, retiring and industrious man," as a writer in *Forest and Garden* (vol. 8, no. 370 (1895), p. 130) described him, is best known for the illustrations reproduced from his delicate and painstakingly detailed drawings. His beautiful watercolors of native trees and flowers, published in chromolithography during his later years, further enhanced his reputation as one of America's greatest botanical artists.

33. A. Hunter Dupree, *Asa Gray, 1810–1888* (Cambridge, Mass.: The Belknap Press of Harvard University Press, 1959), pp. 167–68.

34. Ibid.

35. When the second edition of Emerson's *Report* was published in 1875, a lengthy review by Asa Gray was published in the *American Agriculturist*, vol. 34 (1875), p. 451. Gray stated, "The plates representing the Massachusetts trees and shrubs, are all from the original drawings by Isaac Sprague. When that is said, it is unnecessary to praise them, for in neat and accurate delineation he has no superior. The plates which he contributed to the first edition, here reproduced, are interesting as being almost his first work of the kind."

36. These seven lithographic stones are preserved at the Museum of Amana History. Four have well-preserved images on them; two have damaged images, and one is blank. These stones are evidence that Joseph and Gottlieb Prestele made some of the unsigned plates attributed to them; and that they printed those designs in Iowa. Some, perhaps, were originally issued in Ebenezer, New York. The subjects on these stones are noted in Appendix C, Checklist of Nurserymen's Plates.

37. Prestele to Gray, September 22, 1851.

38. Andrew Denny Rodgers III, *"Noble Fellow," William Starling Sullivant* (New York: G. P. Putnam's Sons, 1940), p. 178.

39. Ibid., p. 170. Prestele engraved all five plates in Vol. 2 (new series), Part I; Plate 2 was printed by Hall & Mooney; Plate 4 by G. & W. Endicott. No printer is given on the other three plates.

Despite the accolades received for the quality of the plates in Part I by Prestele, the five plates in Part II of Sullivant's *Contributions* (published in Volume 4, 1849) were engraved in Boston by Antoine Sonrel, whose services Gray had used on occasion, and were printed by Tappan & Bradford.

40. Through the kindness of Anne F. Clapp, when print and paper conservator at the Winterthur Museum, minute samples of paper from the edges

of Plate 2 and Plate 47 were tested in October 1980. Miss Clapp reported that both samples were of 100 percent rag paper and that they "were found to have protein in their structures, undoubtedly gelatin sizing. Both tested negatively for starch and alum."

41. Prestele to Gray, October 5, 1845; November 21, 1845; December 5, 1845; February 8, 1846. Hall & Mooney, Buffalo, New York; lithographers, engravers, and copperplate printers, were in business from 1839 to 1850. The partners were John P. Hall, who left the firm about 1850, and Lawrence Mooney, who carried on the firm with Charles W. Buell as Mooney & Buell. (Groce and Wallace, op. cit., n. 31.)

42. Prestele to Gray, April 4, 1850.

43. Rodgers, *"Noble Fellow"*, pp. 176–77.

44. Prestele to Gray, December 4, 1846.

45. Notation made by Torrey on Prestele's letter to Torrey of September 26, 1851.

46. "In 1847 the 'great diffusion of horticulture' and the 'munificent patronage' from the public" encouraged the Massachusetts Horticultural Society to plan the issuing of regular volumes of its *Transactions*—including learned papers and news of the society's affairs. (Albert Emerson Benson, *History of the Massachusetts Horticultural Society* [Boston: Printed for the Society, 1929], p. 90.) Early in 1847, Andrew Jackson Downing noted in the *Horticulturist* that the society "is about to publish a very beautiful edition of its Transactions, in a serial form, to be issued once in every two months . . . It will be richly illustrated with coloured plates of fruits and flowering plants, executed in a very superior manner. These plates will be accompanied by authentic descriptions of the varieties represented." In addition, the publication would contain a record of the society's activities and also papers by its members "embracing by far the largest portion of the practical skill and talent in the country." (*The Horticulturist and Journal of Rural Art and Rural Taste* 1, no. 11 (May 1847), pp. 526–27.) The belated first number appeared in large octavo size, containing four chromolithographs (chromos) by William Sharp, thirty-four pages of articles, and twenty-eight pages of proceedings. (Reaction to the Sharp chromos is discussed on pp. 000, and in n. 48.) Downing proclaimed the second number "in the highest degree creditable to the Society. It contains 4 finely colored plates . . . The beautiful manner in which these Transactions are published, certainly equal to the publication of any society on either side of the Atlantic, and the high position which the horticulturists of Massachusetts have attained for practical skill, must, we think, commend these Transactions to general favor." *Horticulturist* III, no. 1 (July 1848), p. 47.

47. William Sharp (b. England 1803) worked in London as a painter, drawing teacher, and, by 1829, as a lithographer. He emigrated to America about 1839, settling in Boston where he advertised as a portraitist. He entered into various partnerships, producing lithographed portraits, botanical prints, sheet music covers, floral designs, and street scenes. One partner, Francis Michelin, had studied chromolithography, which Sharp took up and effectively promoted. (Marzio, *Democratic Art*, pp. 18ff.) More detailed biographical information is in "William Sharp: Accomplished Lithographer," by Bettina A. Norton, in *Art and Commerce: American Prints of the 19th Century*, proceedings of a conference held in the Boston Museum of Fine Arts, May 8–10, 1975 (University of Virginia Press, 1978). See also "Four Chromolithographs of Victoria Regia" by Martia Prather in *The Register of the Museum of Art* (The Helen Foresman Spencer Museum of Art, University of Kansas, Lawrence), vol. 5, no. 8 (Fall 1979), pp. 7–17. A lithographed bust portrait of Sharp is used as the frontispiece to vol. 11 of Hovey's *Fruits of America* (see n. 48 below).

48. "Mr. Hovey's serial," referred to by Downing, was the first installment of the monumental the *Fruits of America* by Charles Mason Hovey (1810–1887), Boston horticulturist, nurseryman, and editor. In his prospectus for the publication, which appears on the back wrap of vol. I, no. 2, dated June 1847, he explained that his new work was to provide fruit growers full information, including colored plates and outline drawings, about the best new varieties of fruit, especially those of American origin, as soon as they had been tested and their quality proven. The colored illustrations would be of paintings made from nature by William Sharp (see n. 47) and "chromolithed and retouched under his direction." The work would be issued in installments in two editions; Royal Octavo at $1 a number, and Imperial Quarto at $2. The first number, promised for January 1, 1847, was to consist of four plates and sixteen pages of text; similar installments would follow every other month until twelve had appeared with a total of forty-eight plates, thus completing vol. 1. Installments for a second volume, of the same size as the first, would then follow.

From the beginning Hovey's plans went awry. The first issue promised for January 1, 1847, was delayed until May 4 because—as the author explained—he needed additional facts. Succeeding issues were to appear every two months, a schedule that proved impossible to maintain. Number 12 of Volume I—with the undated title page for the volume, an index, and a frontispiece portrait of Hovey—was published with the date "February 1850" on the wrap (making it approximately a year late) but may not have been issued until later. Hovey advertised in the *Horticulturist* (vol. 6, May 1851) that the first complete volume of the *Fruits of America* was "now ready" for sale, and that "No. I of volume II, would appear in June (1851)." Volume I was copyrighted in 1851. Reissues of Volume I are dated 1852 and 1853.

Volume II also appeared in twelve installments (lacking dates of issue), resulting in ninety-six pages of letterpress and forty-eight plates. The title page of the completed volume is dated 1856, although it has a copyright notice dated 1851. The index for this volume includes the plates apparently intended for a third volume which was discontinued after sixteen plates and thirty-two pages of letterpress had been issued. (Although 1856, the imprint date of Volume II, is given by bibliographers as the final year of publication, a copy of part 2, Volume III, with "January, 1859" added by hand at an early date, is in the Massachusetts Horticultural Society Library.) Much study remains to be done to sort out the confused record of the publication and even to determine how many plates were actually produced.

Bound copies of the two volumes were offered for sale by Hovey as late as 1866, priced at $35.

Hovey, in the June 1847 issue of his *Magazine of Horticulture*, commented on the production of the colored plates for the *Fruits of America* (Boston, vol. 13, pp. 268–70). "Our artist [Sharp] is too well known to need our praise; but the specimens show that the art of chromo-lithography produces a far more beautiful and correct representation than that of the ordinary lithograph, washed in color, in the usual way. Indeed, the plates have the richness of actual paintings, which could not be executed for ten times the value of a single copy." Hovey gave his strong approval to chromolithography about the same time that the Massachusetts Horticultural Society's Committee of Publication, dissatisfied with the proofs of Sharp's plates for the first issue of the *Transactions*, were anxiously discussing what they could do about that fiasco. Finally, of course, they went ahead and used the plates in the first number of the *Transactions* released in July, about a month after Hovey's statement appeared. Hovey, who was active in the society, must have known the committee's opinion of the Sharp plates.

The criticism of Sharp's plates by both Downing and the publication committee may have been merited, but there are hints that something more than unbiased judgment was involved. Indirectly, of course, they were criticizing Hovey and some of this may have been brought on by the crusty Hovey himself. He had made no mention of the *Transactions* in his *Magazine of Horticulture*, although he regularly reviewed new horticultural publications. Nor, in the first issue of his *Fruits of America* did he refer in any way to the already famous book by Downing, *The Fruits and Fruit Trees of America*, then in its seventh edition, an omission that cannot be excused as an oversight and which—understandably—offended Downing, as did the similarity of Hovey's title to his.

49. Andrew Jackson Downing in the *Horticulturist* 2, no. 6 (December 1847), p. 278.

50. Ibid., p. 287.

51. In 1855, this "Downers Late Cherry" plate was copied in a smaller size, poorly tinted, for an illustration in a gift book, *The Happy Home*. Also used was a poor copy of Prestele's "Bartlett Pear" illustration in the *Horticulturist*, August 1853. Neither illustration identified the artist nor the sources. Rev. A. R. Baker, ed., *The Happy Home*, vol. 1 (Boston, 1855).

52. *The Horticulturist* 3, no. 1 (July 1848), p. 47.

53. Dupree, *Asa Gray*, pp. 189, 453 (n. 10).

54. Gray to Torrey, November 2, 1849, reprinted in Jane Loring Gray, ed., *Letters*, op. cit., n. 26, vol. 1, p. 364.

55. Smithsonian Institution, *Annual Report of the Board of Regents for 1848* (Washington, 1849), p. 157.

56. Smithsonian Institution, *Annual Report . . . for 1849*, (Washington, 1850), p. 171.

57. Senefelder, *Invention of Lithography*, op. cit., n. 28, pp. 149–53.

58. Gray to George Bentham, January 7, 1850, reprinted in J. L. Gray, ed., *Letters*, op. cit., n. 26, vol. 1, p. 366. George Bentham (1800–1884), nephew of the English jurist and philosopher Jeremy Bentham, was a botanist and author of many botanical works.

59. Asa Gray and Joseph Henry, "Articles of a Convention," (ms.), April 15, 1850. Asa Gray Papers, the Gray Herbarium of Harvard University.

60. It is not altogether clear how many subjects were actually referred to in the "Articles of a Convention" because the language is vague and some terms are confusing. The total number appears to be at least fifty-two subjects intended for colored plates, and ten wood engravings from drawings made "under the microscope illustrating the growth and formation of the wood." Eventually, only twenty-three colored plates were actually produced; the fate of the ten woodcuts has not been learned.

The document states that the drawings for eighteen plates had been completed, and twenty-four drawings were in progress. Of the completed drawings, Antoine Sonrel of Boston had "lithographed" (i.e., drawn on stone) two of them and four hundred copies (of each subject?) had been printed (by Tappan & Bradford, Boston; see Appendix A, Checklist of Plates) and sent to Prestele for coloring. Sonrel was engaged in "lithographing" (drawing on stone) two more drawings. Joseph Prestele had completed ten additional plates, some of which he had engraved on stone and some of which he had drawn on stone ("lithographed") "and four reams of plate paper had been forwarded to him in order that he may strike off say, three hundred copies from each for coloring. Four drawings are now in Mr. Prestele's hands for engraving. The drawings of 24 more plates are now in progress, six of which will be finished by the 1st of June." Sprague was to finish only "six or eight or more [drawings] during Dr. Gray's absence in Europe."

61. S. F. Baird, Smithsonian Institution, to Isaac Sprague, December 17, 1850. (Smithsonian Institution Archives, Record Unit 53, vol. 1, p. 122.) Spencer Fullerton Baird, the assistant secretary of the Smithsonian, wrote at the request of Professor Henry, the secretary.

62. Prestele, to Baird, January 27, 1852. (Smithsonian Institution Archives, Record Unit 52, vol. 4, p. 447.)

63. Fritz H. Herrmann, "Zur Lebensgeschichte des Pflanzenmalers Joseph Prestele." *Sonderdruck aus Wetterauer Geschichtsblätter.* (Friedberg/Hessen, 1979), vol. 28.

64. Prestele to Gray, January 5, 1850; February 16, 1850; June 21, 1854.

65. The notation, on a plate in the library of the Hunt Institute for Botanical Documentation, Carnegie-Mellon University, Pittsburgh, Penn., is followed by a second, partially erased, in another hand. (Bernadette G. Callery, librarian, to the author, June 29, 1982). In 1853, Prestele explained to Dr. Torrey that many of the drawings made on the government expeditions to the Far West were so poorly done that he had to redraw them before they could be engraved.

66. E. Foreman, Smithsonian Institution, to Isaac Sprague, April 28, 1851, Gray Papers, Gray Herbarium of Harvard University.

67. Prestele to Gray, December 31, 1852, with notation by Gray. (Smithsonian Institution Archives, Record Unit 52, vol. 4, p. 435.)

68. Prestele to Gray, January 5, 1850.

69. Antoine Sonrel, an artist and engraver about whom little seems to have been published, was working in the Boston area c. 1853—after 1862. He copied Sprague's drawings on stone for Plates 1, 3, and 4 for the projected *Forest Trees of North America.* These were printed by Tappan & Bradford. (Groce and Wallace, op. cit., n. 31, p. 593.) Sonrel also made drawings for illustrations in Thaddeus W. Harris's *A Treatise on Some of the Insects Injurious to Vegetation* (Boston, 1862), p. iv.

70. A detailed history of the development of chromolithography is given by Peter C. Marzio, *Democratic Art*, op. cit., n. 31.

71. The Smithsonian Institution, *Annual Report . . . for 1850*, p. 186.

72. The Smithsonian Institution, *Annual Report . . . for 1851*, p. 15.

73. Dupree, op. cit., n. 33, p. 168.

74. Prestele to Gray, November 13, 1855. Prestele also says that he may be instructed to leave for Iowa soon and asks Gray to let him know what "is to be done with the stones and paper that is left here in my house."

75. The Smithsonian Institution, *Annual Report . . . for 1856*, pp. 32–33.

76. On December 2, 1850, Prestele wrote Sprague that he had forwarded to him by railroad freight a box containing "1600 coloured plates," itemized as: Plate 35, 350 copies; Plate 49, 300 copies; Plate 30, 100 copies; Plate 50, 100 copies; Plate 39, 150 copies; Plate 46, 100 copies; Plate 31, 100 copies; Plate 47, 150 copies; Plate 4, 250 copies. He itemized his second shipment of 1,329 plates on February 11, 1851, as containing: Plate 3, 390 copies; Plate 4, 139 copies; Plate 35, 50 copies; Plate 46, 100 copies; Plate 50, 100 copies; Plate 31, 175 copies; Plate 30, 175 copies; Plate 49, 100 copies; Plate 39, 100 copies.

77. Tappan & Bradford, engravers and lithographers, operated in Boston 1849–54. The partners were the engraver Ebenezer Tappan (1815–1854) and the lithographer Lodowick H. Bradford (b. Massachusetts c. 1815). Following Tappan's death in 1854, Bradford headed the firm of L. H. Bradford & Co. 1854–59. (Groce & Wallace, op. cit., n. 31, p. 75.)

78. The Smithsonian Institution, *Annual Report . . . to July, 1891*, p. 60.

79. Prestele's signature is on these four plates (nos. 5a, 14, 71, and 75), appearing both as "J. Prestele" and "J. Prestile [*sic*], Del." They were "Aqua-tinted and Printed by J. I. and R. H. Pease." Joseph I. Pease (1809–1883) was a line engraver, watercolor and crayon artist, and the older brother of the lithographer, Richard H. Pease (1811–1869). Joseph worked briefly with Richard while he was located in Albany (1834) and in Philadelphia (1835). Richard remained in Philadelphia until 1850. (Groce and Wallace, op. cit., n. 31; and Katharine McClinton, "American Flower Lithographs," *The magazine Antiques* 49 (June 1946), pp. 361–65.)

80. The illustration for the two-page article is signed: "Engraved by J. Prestele. Lith. of Wm. Endicott & Co., N. York." Gray, in his review of the article, mentions that the drawing was made by Sprague (*American Journal of Science and Arts*, series II, vol. 16, 1853, p. 426.)

81. Prestele to Torrey, September 26, 1851. By November 24, 1851, he had sent Torrey proofs.

82. Torrey's statement, dated New York, January 12, 1857, is in the introductory section of Torrey's "Description of the General Botanical Collections," in Lieutenant Amiel Weeks Whipple's *Exploration and Surveys for a Railroad Route . . . in 1853–54* published in volume IV of the *Reports of Explorations and Surveys to Ascertain the Most Practicable and Economical Route for a Railroad from the Mississippi River to the Pacific Ocean. Made under the Direction of the Secretary of War, in 1853–(56).* See Appendix B for the full titles of the Army reports mentioned in the text.

83. Prestele to George Engelmann, December 4,

1857. (George Engelmann Papers, Correspondence, Missouri Botanical Garden Library, St. Louis, Mo.)

84. James Ackerman (b. New York c. 1813) was a partner in Ackerman & Miller, sign and banner painters in New York City 1838–65. He exhibited colored lithographs at the American Institute in 1845 and 1847 (Groce and Wallace, op. cit., n. 31). He also produced colored lithographs of flowers for book illustrations, including those in *Flora's Lexicon; An Interpretation of the Language and Sentiment of Flowers* (Philadelphia: P. S. Duval, 1839). Later he took up chromolithography. He printed thirty-one of the chromos and four uncolored lithographs for Henry R. Schoolcraft's *Historical and Statistical Information Respecting the History, Condition, and Prospects of the Indian Tribes of the United States*. When volume I appeared in 1851, the Ackerman plates were criticized for failing to do justice to the original watercolors by Seth Eastman, the primary illustrator for the book. (Marzio, *Democratic Art, op. cit.*, n. 31, pp. 29–30.)

Prestele's relations with Ackerman are suggested by comments in the artist's letters to John Torrey. December 28, 1853: "I received nothing yet from Mr. Ackerman. I am sorry that you have so much trouble with him concerning my bill." May 15, 1856: "I have received today a letter of A. G. Seaman of Washington in which he stated [to] me that my proposal is accepted for my engraving the Botanical Plates to $10 each, and that I may procure myself with stones necessary from James Ackerman of New York. I suppose you will have sent my bill now to him. Please let me know . . . if it is necessary that I also shall sent [*sic*] him one or not." February 4, 1857: "I shall send also immediately the stone to Mr. Ackerman for printing."

85. Prestele, Ebenezer, New York, to Christian Metz, at Amana, Iowa, April 10, 1857. Quoted in *Inspirations-Historie*, volume for the years 1817–67, p. 5 of the extracts supplied the author by Henrietta M. Ruff, Amana Heritage Society, 1980, op. cit., chap. 1, n. 10. Translated by Elisabeth Kottenhahn.

86. For biographies of James Vick see: L. H. Bailey, *Cyclopedia of American Horticulture* (New York: The Macmillan Co., 1902), vol. 4, p. 1928; and the *Dictionary of American Biography*. Although the evidence is inconclusive, it appears that Vick was one of the first to use color plates in seed catalogues, which he did around 1864.

87. *The Genesee Farmer*, vol. 14, no. 1 (January 1853), p. 40.

88. (William Henry Prestele). *187[9]. Catalogue of Fruit and Flower Plates Drawn from Nature. Lithographed and Colored by Wm. H. Prestele, Son of the Late Joseph Prestele, Sen., of Amana Society, Homestead, Iowa. Iowa City, Johnson Co., Iowa.* Eight pages; colored frontispiece and an uncolored plate of fruit; wraps, printed cover, 23.8 × 15 cm. (Author's collection.)

89. The design of Joseph Prestele's "Beurre D'Aremberg" pear appeared in the Massachusetts Horticultural Society's *Transactions* 1, no. 3, which, after a long delay, was issued under the date of January 1852. This design was also published with minor changes as a nurserymen's plate; one variant has the Amana imprint. (See Checklist of Nurserymen's Plates, Appendix C.)

90. Prestele's reference to making nurserymen's plates is in his letter to George Engelmann, December 4, 1857, op. cit., n. 83.

91. Many of the nursery's account books, its office library (or the considerable part that remains of it), and other data are also preserved in the Ellwanger and Barry Collection, Department of Rare Books and Special Collections, Rush Rhees Library, The University of Rochester.

92. George Ellwanger was born in Gross-Heppach, "in the Remsthal, called 'the garden of the fatherland,'" the kingdom of Württemberg, Germany, December 2, 1816, and died in Rochester, New York, in November 1906. Biographical information about him seems surprisingly scant—evidence, perhaps, of his innate modesty. His brief autobiography, together with a sketch of Patrick Barry's life, and the history of the Ellwanger & Barry nursery, is in *The Garden of the Genesee*, published by the nursery in 1940, its centennial year. (It is an illustrated 29-page pamphlet, without editor or publisher's imprint.) Informative accounts of the Mount Hope Nursery and its proprietors, Ellwanger and Barry, are given in the following articles in the University of Rochester *Library Bulletin*, vol. 35 (1982): "From the Genessee to the World," by Diane Holahan Grosso; "A Family Story: The Ellwangers and the Barrys," by Alma Burner Creek; "The American Dream on Mt. Hope: Nineteenth-Century Building by Ellwanger & Barry," by Susan Sutton Smith; and "A Business Library for Ellwanger and Barry," by Alma Burner Creek. See also: Blake McKelvey's "The Flower City; Center of Nurseries and Fruit Orchards," in the *Rochester Historical Society Publications*, vol. 18 (1940), pp. 121–69; William F. Peck, *History of Rochester & Monroe County* (New York and Chicago: Pioneer Publishing Co., 1908), pp. 1173–74.

More information is available about Patrick Barry (b. Belfast, Ireland, May 24, 1816; d. June 23, 1890). See: *The Dictionary of American Biography*; *The American Agriculturist*, vol. 46 (1887), p. 461; *Garden and Forest*, vol. 3 (1890), p. 328; William F. Peck, *History of Rochester*, pp. 1260–69, and the University of Rochester *Library Bulletin*, and L. H. Bailey, *Cyclopedia of American Agriculture*. (New York: The Macmillan Co., 1909), vol. 4, p. 554.

93. The "Summer Rose" plate is in Box 35a, Folder 4, Ellwanger and Barry Collection, University of

Rochester. Also there is a plate of the "Andrews Pear" with the penciled notation, "A little more yellow."

94. *The Rose* (New York: Dodd, Mead & Company, 1892), p. 31.

95. See n. 91. The account books record the amounts of the purchases made yearly from 1854 until 1860, first from Joseph Prestele and then from the Amana Society. Unfortunately, little of the office correspondence has survived, and details of the orders from Prestele have not been found. The unbound plates by Prestele are on folio sheets of unwatermarked rag paper, about 34 × 26.5 cm. The bound examples are smaller, having lost several inches in both directions from trimming, which also removed any signatures that may have been on the plates.

96. Thirty-nine plates of apples, pears, plums, roses, and flowering trees and shrubs, which were either signed by Joseph or can be attributed to him or Gottlieb Prestele, are in a bound assortment of seventy-three plates with the cover title, *Album of Fruit / Ellwanger & Barry / Mt. Hope Nurseries / Rochester, N.Y.* Inside the front cover is a partially printed label" *F. W. Kelsey* [handwritten] / Agent for Ellwanger & Barry / Mount Hope Nurseries, / Rochester, N.Y. / *Spring* 18 *71* [handwritten]." (Author's collection.)

Fred W. Kelsey, to whom the volume was given in 1871 as an agent for Ellwanger & Barry, founded his own nursery business several years later. In his 1877 catalogue he explained that he raised no stock but served as agent for his customers, carefully selecting for them the plants they ordered.

Three of the plates in the Ellwanger & Barry plate book have Joseph Prestele's signature. As noted in the text, the eccentric form of the letters in the lithographed captions, the variations in their size, and the uncertain lines redrawn make the inscriptions on the plates distinctive. Possibly other plates in this collection were also published with the Presteles' signatures, although that too is uncertain. The artists were inconsistent in the spacing they used between their captions and signatures (1.5, 2.0, and 2.2 cm. being found), so that any signatures that may have been on some plates were trimmed off in binding. None of the plates carry the imprints of nursery plate dealers or publishers.

A bound assortment of nurserymen's plates similar to the volume discussed above is in the Ellwanger and Barry Collection at the University of Rochester. Stamped on its cover is: *Album of Fruit / Ellwanger and Barry / Mt. Hope Nurseries / Rochester, N.Y.* with a penciled notation, "Sample book for office." It contains sixty-six plates, 32 × 24 cm. Six are signed by either Joseph Prestele or his son Gottlieb while at Ebenezer, which dates them before 1858; others are unsigned but can be attributed to them.

It is impossible to date the Prestele plates in these two books with any certainty. They seem to have been issued at different times. Some are different states of the Prestele engravings used in the *Horticulturist* 1853–55. Which came first?

97. One of the first people to dabble in the plate business after Joseph Prestele had begun was the restless Thomas Wright who seems to have bought his plates from publishers (Prestele?) and retailed them. He was a sometime "nurseryman" living in Syracuse, New York, in the autumn of 1859, when his list of "Colored Plates of Fruit for Nurserymen," was mentioned by Thomas Meehan, Philadelphia, in his *Gardener's Monthly*, vol. 1, no. 10 (October 1, 1859), p. 157. Could he have issued earlier lists?

Wright first appeared in the Syracuse city directory for 1859–60 as "Thomas Wright, Nurseryman, h. Russel n. Shonnard." No business address or business association is given. Wright next turned up in Rochester where, in October 1860, he advertised: "Colored Fruit Plates. None but the best for sale by Thomas Wright, Nurseryman, Rochester, New York. (Formerly of Syracuse.) Price $35 per hundred." (*Horticultural Advertiser*, appended to *The Gardener's Monthly* (Philadelphia), vol. 2 (October 1860), p. 4. In the Rochester directory for 1863–64, Thomas Wright is listed as "Nurseryman"; in 1864–65 as a "Commission Tree Broker, 35 Reynold's Arcade, h.Henrietta." He was not listed in the 1870 Rochester directory. See also: Karl Sanford Kabelac, "Nineteenth-Century Rochester Fruit and Flower Plates," University of Rochester *Library Bulletin*, op. cit., n. 92, pp. 93–113. *The Gardener's Monthly and Horticultural Advertiser* (Philadelphia), vol. 1, no. 10 (October 1, 1859), p. 157.

An early publisher of plates was Erastus Darrow (b. Plymouth, Connecticut, January 29, 1823: d. Rochester, New York, March 21, 1909), and his brother Wallace Darrow, who were partners 1856–66, as E. Darrow & Brother. They were "Wholesale and Retail Booksellers, Stationers & Publishers . . . And Publishers of Fruit and Flower Pictures for Nurserymen . . . 65 Main St. Rochester." (Rochester city directory for 1861, p. 113). Henry Kempshall was Erastus's partner from 1866 until Kempshall's death in 1868. Thereafter Erastus was the sole owner of his large book and stationery firm, which he headed for many years. He was married in 1846 to Miss Susan R. Martin (d. 1871). After her death he married Miss Sophia C. Munger in 1877. He died aged eighty-six, survived by his wife, children, and grandchildren. (*Rochester Democrat and Chronicle*, March 22, 1909.) See also Kabelac, "Nineteenth-Century Rochester Fruit and Flower Plates," op. cit., n. 92, p. 97.

98. The Kellogg firm originated with D. W. Kellogg in Hartford, Connecticut, about 1828, who operated it until 1842 when two other brothers took it over. The firm was then renamed E. B. & E. C. Kellogg. From 1848 to 1850 the firm was Kellogg & Comstock. During 1851–55 the two brothers operated independently, but resumed their

partnership in the latter year, continuing until 1867, when the firm of Kellogg & Bulkeley was formed. By 1850 the Kelloggs were doing a very large business. They made nurserymen's plates for E. Darrow & Brother, Rochester, as well as for D. M. Dewey. (Frances Phipps, "Connecticut's Printmakers: the Kelloggs of Hartford," *Connecticut Antiquarian* 21, no. 1 (1969), pp. 19–26; Groce and Wallace, op. cit., n. 31, p. 363.

99. The Kellogg plates were clearly intended as nurserymen's plates. They are not the sort of decorative and ornate prints designed for framing, such as various nineteenth century lithographers published, including the Kelloggs. Their fruit plates (the only subjects located) picture a single specimen with stem and leaves, designed in the manner that Joseph Prestele used for his nurserymen's plates, a not surprising fact since a number of the Kellogg fruit pieces appear to have originated with Prestele.

100. John Sage and his sons, Henry H., John B., and William, were lithographers and music dealers in Buffalo, New York, 1856–60. (Groce and Wallace, op. cit., n. 31, p. 554.

101. On December 11, 1856, Joseph Prestele wrote from Ebenezer to Asa Gray, "I suppose you will have seen a young artist by the name of Mr. A. Hochstein, whose father lives in my vicinity and is an esteemed friend of mine. I was informed that he was recommended by a Lady of Rochester, N.Y. He can make drawings pretty well, & I would beg your kindness to furnish him with a good place." Gray sent the information to Torrey, who was then in much need of draftsmen to make drawings for reports of western expeditions, and who employed Hochstein. (Andrew D. Rodgers III, *John Torrey* [London & Oxford: Princeton University Press, 1942], pp. 227, 247.) In 1861 the editor of the *Horticulturist* (vol. 16, pp. 45–46) stated that Hochstein had painted fruit and flower pieces used in the magazine and that he considered Hochstein, in his particular line, one of the best artists in the country. Most of the plates in the 1861 volume are his work; several of the plates are signed "A. Hochstein." The plates were published by C. M. Saxton, New York City. Groce and Wallace (op. cit., n. 31) mention that Hochstein was a miniature painter of New York City in 1859 and that he exhibited at the National Academy 1868–72.

102. George Frauenberger (b. Hildburgshausen, Germany, 1829; d. Rochester, New York, April 18, 1899), "Wood Engraver," advertised in the Rochester city directory, 1861, that he was a "Designer and engraver on wood . . . prepared to make Drawings and designs of Machinery, &c., for Applicants for Patents. Also, draw from life and engrave portraits . . . Landscapes, Fruits, Outlines, &c." In 1878, D. M. Dewey advertised colored engraved prints by Frauenberger of some sixty varieties of "Trees, Shrubs, Flowers, &c." priced at from $3 for 100 of one kind to $20 for 1,000 different plates. Dewey also advertised Frauenberger's engravings for illustrating gardening catalogues and periodicals. Examples of Frauenberger's work appeared in the catalogues of nurserymen and seedsmen and in periodicals in western New York and elsewhere. (See Kabelac, "Nineteenth-Century Rochester Fruit and Flower Plates," op. cit., n. 92, p. 98.)

103. These lists are mentioned in the *Gardener's Monthly* (Philadelphia), vol. 1 (1859), pp. 42, 122.

104. The complete title page reads "*The Colored Fruit Book, for the Use of Nurserymen, Containing Accurate Specimens of Colored Fruits and Flowers, Carefully Drawn and Colored from Nature, and Designed to Represent a Medium and Fair Size of Each Particular Fruit. It has been the desire and design of the Publisher of this Series of Colored Fruit Plates, to place before the purchaser of Fruit Trees, as faithful a representation of the Fruit as is possible to make by the process adopted.* Rochester, N.Y. 1859. Published by D. M. Dewey, Horticultural Bookseller. Catalogues of over 200 varieties of colored fruits and flowers, (of which these are samples), with prices, will be furnished on application by mail to the above address." (Author's collection.)

A similar collection, also titled *The Colored Fruit Book*, and with an only slightly reworded title page, was published in 1862. It too was bound with a price list of plates which, as Dewey stated, had then grown to "over 400 Varieties." Embossed on the cover is "Dewey's Colored Fruits & Flowers. / M. D. Freer & Co." (Local History Room, Rochester Public Library, Rochester, New York; courtesy of Wayne Arnold.)

105. D. M. Dewey, *The Tree Agents' Private Guide.* (Rochester, N.Y.: D. M. Dewey, 1875.) A copy is in the Ellwanger & Barry Collection, University of Rochester Library. (See n. 110 following.)

106. Ibid., p. 12. Dewey also explained in *The Tree Agents' Private Guide* that for the first plates he had designs "drawn for him by competent artists; had about 150 lithographic stones made from which impressions were taken and colored by hand. He also made about 150 other varieties which were colored entirely by hand. With these 300 varieties he issued his first catalogue for nurserymen.

107. D. M. Dewey and his nurserymen's plates are discussed and pictured in two works by the author: *A Nineteenth Century Garden* (New York: Main Street Press / Universe Books, 1977); and "Drawn and Colored from Nature. Painted Nurserymen's Plates." (The magazine *Antiques* 73, no. 3 (March 1983), pp. 594–99.

108. One of the defectors was Daniel W. Sargent who had worked for Dewey as a clerk from 1864 to 1869, then left to enter the plate business for himself. In 1871 he advertised for sale a stock of 1,000 varieties of fruit and flower plates.

Sargent's venture seems to have been short-lived. One of his theorem and freehand plates—"Glory of the Mosses"—is in a bound collection of medium-sized plates with a Dewey title page, c. 1880. The Sargent plate is smaller in size than those of Dewey's, being 21.5 × 13.5 cm. untrimmed. The quality is poor. In the Ellwanger & Barry Collection, University of Rochester, is a theorem and freehand plate, 29 × 23 cm., "Double-Flowering Rocket. (Hesperis m. fl. pl.) D. W. Sargent, Rochester, N.Y. Fruit and Flower Plates, 1000 Varieties." Sargent (1823–1911) is buried in Mt. Hope Cemetery, Rochester. (Kabelac, op. cit., n. 92, p. 102.)

109. It is not clear when Dewey first introduced the smaller plates. He continued selling "hand-made" plates in his original size (11 by 9 in.; 28 × 23 cm.) until he retired from business in 1888. His *Dewey's New Catalogue of Colored Plates of Fruits, Flowers, Etc.* (1871), listing "over 1500" subjects, does not mention plate size, suggesting that the smaller sizes had not been introduced. But by the following year "Dewey's Pocket Series" (22 × 14.5 cm.) was being published as some plates with that imprint were copyrighted in 1872. In his *Tree Agents' Private Guide* (op. cit., n. 105, p. 14), Dewey mentions "the new pocket edition" which he introduced in 1873 and "which meets with the general approval of agents, especially those who are canvassing in large cities." See also *No. 4. Dewey's Classified Catalogue* (1878). (A copy of Dewey's 1871 *New Catalogue of Colored Plates* is in the Local History Room, Rochester Public Library. In design it is a four-page circular, about 35 × 20.5 cm.)

By 1875 Dewey claimed to have a stock of from 100,000 to 200,000 plates, representing over 2,400 varieties "of the most popular, large and small fruits, flowers, shrubs, ornamental trees, etc., grown by nurserymen in the United States and Canada, besides a series of plates illustrative of designs and suggestions for landscape gardening." He also found that deluxe versions of his plate books—gilt-edged, bound in morocco or calf, and with decoratively embossed covers—were popular among horticultural societies, for prizes at fairs, and for parlor tables alongside the family Bible and photographic album. They were priced at from $10 to $50, and their elegance proclaimed their cost. One example—*The Specimen Book of Fruits, Flowers and Ornamental Trees . . . Published by D. M. Dewey, Ag't.*—has a beautifully lithographed title page ornamented with a border of fruits and flowers, delicately tinted. "John Cooper" is embossed in gold on its cover. The book is undated but internal evidence shows that it was issued after 1867.

That specimen book includes a large number of plates, many of flowers. Some plates are examples of those first issued by Dewey, others are later, a few are poorly chromolithographed plates by E. B. & E. C. Kellogg, and the successor firm (1867) of Kellogg & Buckeley. (Local History Room, Rochester Public Library.)

Dewey advertised his "Book of Plates, in extra gilt binding, suitable for parlor tables, or for premiums for Horticultural Societies," in his *New Catalogue of Colored Plates* (1871), see above. See also *No. 4 Dewey's Classified Catalogue of Colored Fruit Plates* (1878). Another handsome edition of a Dewey plate book (c. 1871) with fifty-one plates is in the Local History Room of the Rochester Public Library. Embossed in gold on its black cloth cover is *Nurserymen's Colored Fruit & Flowers*, and "F. E. Kortnight, Stroudsburg, Pa.," apparently the name of the original owner. The full title given on the title page is *The Nurseryman's Specimen Book of American Horticulture and Floriculture. Fruits, Flowers, Ornamental Trees, Shrubs, Roses, &c., Colored from Nature.* The flyleaves are marbelized. There is no date or imprint. In addition to the plates there is a blank page with the heading "Special List of Fruit and Ornamental Trees, Shrubs, Roses, Flowers, Etc. for Sale. Specimens of which are not shown in this book." A similar volume with sixty-six plates measuring about 27 × 21 cm. is in the Rare Book Room of the Huntington Library.

110. To supply plates for his expanding business, Dewey employed an average of thirty persons, including "several skilled German, English, and American artists." He added "Nurserymen's requisites" of tools, labels, and other supplies to his sales items and expanded his list of cheap colored wood engravings by George Frauenberger. Then, to further promote his business, he wrote and published a handbook on salesmanship for the benefit of the traveling nursery agents. *The Tree Agents' Private Guide* (1875) is a little book of seventy-eight pages, of pocket size with a foldover flap and clasp, resembling an old-fashioned wallet. In it Dewey sought to cover every possible situation any tree agent might face. (See n. 105; a copy of the *Guide* is in the Ellwanger & Barry Collection, University of Rochester Library.)

111. Dewey's last catalogue, *Dewey's Classified Catalogue Number Eleven* was published in the *Horticultural Magazine. An Illustrated Monthly Devoted to the Interests of all Engaged in Developing and Disseminating Fruits of the Earth* (Rochester Lithographing & Printing Co., Publishers, Rochester, N.Y.), vol. 1, no. 5 (May 1888), pp. 1–(46?). It lists about 3,000 plates. With it came his announcement that he had "consolidated" his fruit plate and nursery supply business with that of the Rochester Lithographing and Printing Company. He died the following year.

112. To my knowledge, the first writer to call attention to D. M. Dewey and his nurserymen's plates was Carl W. Drepperd in his article "The Tree, Fruit and Flower Prints of D. M. Dewey, Rochester, N.Y. from 1844. A First Essay at Definitive Listing of his Often Gorgeous Prints Done by Various Lithographic Processes and Stencil Coloring" (*The Spinning Wheel*, May 1956), pp.

12 ff. Included in his checklist of 134 of the many plates issued by Dewey (which list the author explained was not exhaustive) was the "precise title given with all other lettering." Drepperd (incorrectly?) states that Dewey began issuing his plates in 1844 but he did not give his source for that statement.

Chapter 3

1. Bertha M. Shambaugh. *Amana That Was and Amana That Is* (New York: Arno Press, 1976; first published 1932), p. 66, See chap. 1, n. 15.

2. Frank J. Lankes, *The Ebenezer Society*. Reprint of 1949 edition. (West Seneca, New York: West Seneca Historical Society, 1963), pp. 122–23. See chap. 2, n. 1.

3. Shambaugh, *Amana*, p. 88.

4. Ibid., p. 87.

5. Ibid., p. 89.

6. The signatures "By J. Prestele. Amana, Iowa Co., Iowa"; "By J. & G. Prestele, Amana, Iowa"; and "Amana Society by I. & G. Prestele, Amana, Iowa County, Iowa" may have been among the first used by Joseph Prestele, Sr., and Gottlieb Prestele at Amana. Gottlieb's plates have "Drawn from Nature & Colored by G. Prestele." Then came "Lith. & col[d.] by Amana Society, Iowa County Iowa" or such variations as "Lith. & Sold by Amana Society. Amana, Iowa County, Iowa." At Amana their work continued to be of high quality, and the size of their plates remained at 34 × 27.7 cm. untrimmed.

7. The last entry in the "Joseph Prestele" account was on March 23, 1860, as stated earlier. Some of the firm's account books appear to be missing from the Ellwanger & Barry Collection at the University of Rochester Library. All but a few orders and letters, and all of the invoices are lacking, apparently destroyed before the university received the collection. See chap. 2, nn. 91, 95.

8. Efforts to learn what "plates at the Agricultural Department at Washington" were "coloured" by the Presteles have been unsuccessful. The department was created by act of Congress approved May 15, 1862. Annual *Reports* by the commissioner of agriculture for the years 1862, 1863, and 1864 do not include any colored plates, nor did such colored plates appear in the annual reports until much later. (The quotation from the *Muscatine Weekly Journal* was kindly supplied by Karen Laughlin, librarian, Iowa State Historical Society, Iowa City, Iowa.)

9. The Joseph Prestele *List of Fruit-and Flower-Plates* (c. 1860) is included with 180 unbound plates, approximately 34 × 26.5 cm., in the History of Science Collections, Cornell University Libraries. (Three additional black-and-white engravings of fruit, not by the Presteles, joined the collection at some point.) Most of the plates are without imprint or signature. About forty-five had the imprint "Amana Society, Iowa County, Iowa." Three were signed by Gottlieb Prestele, one by Joseph and Gottlieb Prestele; and one by Joseph Prestele with his address, "Ebenezer, n. Buffalo." (Information courtesy of Daniel W. Corson, librarian, History of Science Collections, Cornell University Libraries, August 6, 1981, to the author.)

10. *Inspirations-Historie*, Amana Heritage Society, Amana, Iowa, p. 875. Copies by Henrietta M. Ruff, and letters of Mrs. Ruff to the author, May 13, and July 14, 1980. Translated by Elisabeth Kottenhahn. Op. cit., chap. 1, n. 10.

11. Ibid., p. 187–88.

12. Fritz H. Herrmann, "Zur Lebensgeschichte des Pflanzenmalers Joseph Prestele," p. 155, op. cit., chap. 1, n. 3, quoting from Claus Nissen, *Die botanische Buchillustration*, op. cit., chap. 1, n. 2.

Chapter 4

1. Fritz H. Herrmann, "Zur Lebensgeschichte des Pflanzenmalers Joseph Prestele," *Sonderdruck aus Wetterauer Geschichtsblätter*, vol. 28 (Friedberg/Hessen, 1979), p. 140. Translation by Elisabeth Kottenhahn.

2. Joseph Prestele, Sr., Ebenezer, New York, to his family in Germany, October 15, 1843. This was the first letter that Joseph, Sr., wrote to his family after his arrival in America. (Courtesy, Amana Heritage Society, translation by Magdalena Schuerer, Amana, Iowa. See chap. 2, n. 4.)

3. Joseph, Jr.'s wife and daughter are mentioned in the U.S. Census of 1870 for McLean County, Illinois. A check of the New York City directories, 1850–67, produced entries for two Joseph Presteles. One in 1853–54 was a locksmith; the second, in 1863–64, a machinist. Presumably, neither of these was the artist. Mrs. Joseph Mattes, a Prestele descendant, Amana, Iowa, in 1983 said she believed that Joseph Prestele, Jr., had only one child, Elisabeth, who never married, and lived in New York until the mid-1920s.

4. Data supplied by Greg Koos, archivist, McLean County Historical Society, Bloomington, Illinois.

Chapter 5

1. Prestele to Torrey, April 7, 1845. See chap. 2, n. 3.

2. Prestele to Gray, September 22, 1851. Gray had just returned from Europe and was dissatisfied with some of the work that Prestele had done while he was gone. Prestele, much concerned, wrote to offer his apologies. "I feel great grief that some of the plates, which I engraved and colored since your absence do not please you at all, and I would very much regret if this would be the cause to abandon the work." He went on to say that while Gray was

in Europe, he had sought to speed up production by having some of the coloring done in New York [by Joseph, Jr.?] but he was disappointed in the results. The recent shipment consisted of plates colored "mostly by the hands of my son [Gottlieb?] and daughters, which I believe will please you somewhat more." He then proceeded to ask if Gray could find work for Gottlieb during the winter.

3. A copy of Gottlieb Prestele's *Specimen Book of Fruits & Flowers* is in the Library of the Ohio State University (Department of Special Collections), Columbus, Ohio. Originally the collection of plates with which it is bound contained twenty-five plates, as shown by its handwritten list, but one is now missing. (Information courtesy of Robert A. Tibbetts, curator of special collections, July 20, 1981, to the author.)

4. The file copies of Gottlieb's plates are in the collections of the Amana Heritage Society, Museum of Amana History.

5. A copy of this undated, one page (broadside) *Catalogue* is in the History of Science Collections, Cornell University Libraries, courtesy of David W. Corson, history of science librarian.

6. A copy of the last *Catalogue* (undated, ca. 1875, a one-page broadside) is in the collections of the Amana Heritage Society, Museum of Amana History, and also in one or more private collections in the Amanas.

7. Mentioned in the *Inspirations-Historie*, a copy of which is in the collections of the Amana Heritage Society. Transcription supplied by Henrietta M. Ruff; translated by Elisabeth Kottenhahn, 1980, pp. 187–88. "The printing patterns for the calico are not only designed but made by a member of the Society. The colors used in the dyeing are chiefly blue, brown, or black. Only the severest of these patterns are considered sufficiently 'plain' for the dress of the Amana women and children." (Bertha M. H. Shambaugh, *Amana That Was and Amana That Is* [New York: Arno Press, 1976; first published 1932, pp. 143, 344].)

8. Gottlieb's paint box is in the Museum of Amana History.

Chapter 6

1. The U.S. Census of 1870 for McLean County, Illinois, gives "W. H. Presley [*sic*], 31, Born in Prussia, a painter of fruits and flowers." The 1880 Census of Lucas Township, Johnson County, Iowa states that Prestele was born in Hessen-Darmstadt and was then forty-two years old. By the time William Henry was born on October 13, 1838 (see chap. 1, n. 11) his parents had moved from Bavaria with other Inspirationists to the property of the former convent of Engelthal, Hessen-Darmstadt, which the Community had leased.

2. Henrietta M. Ruff, Amana Heritage Society, to the author, February 16, 1981, quoting from the *Inspirations-Historie*, testimony no. 10. Metz's admonition was given in Middle Ebenezer, February 21, 1854.

3. The Museum of Amana History; Dr. Louis A. Clemens, Homestead, Iowa.

4. The U. S. Census of 1870 does not give Ann's maiden name. Their daughter's name is given as Margaret in this census, but the U.S. Census of 1880 shows a daughter named Elisabeth, with the same birth date.

5. Franklin Kelsey Phoenix was born March 3, 1825, in the town of Perry, Wyoming County, New York, the son of Samuel F. and Sarah A. Phoenix. By 1870, when his nursery was described as the second largest in the nation (Bloomington, Illinois *Weekly Pantagraph*, March 30, 1870), Phoenix was a young-looking forty-five, about 5 feet 8 inches in height. Seven years later, in 1877, his empire crashed; his production of nurserymen's plates may have ceased a year or so earlier. It was a "deeply mortifying failure," as he later commented without suggesting the cause. (John H. Burnham, "Early Nurseries," *Transactions of the McLean County Historical Society*, Bloomington, Illinois, vol. II, 1903, p. 313, quoting from a letter of F. K. Phoenix, 1893. His biographers did not provide an explanation either. A possible clue is in a newspaperman's comment that Phoenix was "somewhat impulsive." ("The Bloomington Nursery" in the Bloomington, Illinois *Weekly Pantagraph*, March 30, 1870.) After his debacle, Phoenix returned to his childhood home at Delavan, Wisconsin, where he died in 1911, greatly honored for his good works.

Although the Bloomington Nursery passed into the hands of creditors and underwent various changes in ownership (in 1894, the wholesale catalogue reported that the nursery was then owned by the Phoenix Nursery Company, successors to Sidney Tuttle & Co.), it continued operations for many years on a modest scale. Nurserymen's plates continued to be advertised in its catalogues. As the original stock was sold off, plates by Dewey and other publishers were substituted. The wholesale catalogue for the spring of 1886 advertised "Colored Plates of Fruits and Flowers, drawn and colored from nature . . . Our Plates are Thompson's best—all new, handsome and showy. 5½ by 9 inches, with printed descriptions at bottom of each Plate . . . Price, 14 cents per Plate." The wholesale catalogue for Spring 1894 announced "We have now secured the exclusive Western Agency for the celebrated Thompson Hand-Painted Plates made by the Nicholson Company.") Prices were steadily reduced as the market declined; in 1894 hand-colored plates could be had for 8 cents each, "binding extra."

For additional biographical information on Phoenix see: John H. Burnham, "Early Nurseries," pp. 310–15; "Pioneer Citizen Hears Death's Summons," *Delavan* (Wisconsin) *Republican*, February 9, 1911; "Bloomington Nursery and Garden," in *Evergreen City*, R. S. Lawrence Comp. (Bloomington, Illinois: R. S. Lawrence, 1871), p. 60; *Holland's Bloomington City Directory, for 1870–71* (Chicago: Western Publishing Co., n.d.) p. 35.

6. *Holland's Bloomington City Directory, for 1870–71*, p. 35.

7. Burnham, "Early Nurseries," op. cit., n. 5, p. 312.

8. Bloomington, Illinois *Weekly Pantagraph*, March 30, 1870.

9. A "William Bristely [Prestele?], laborer" is given in the 1868–69 Bloomington directory. The dates I have used related to the Presteles' stay in Bloomington are based upon listings in the city directories, from the directory for 1868–69, to the directory for 1874–75, and reflect my assumption that the information in them was compiled the year before the publication date.

10. This bound assortment consists of fifty-four plates (trimmed to 27.5 × 21 cm.) of which twenty-four plates were published by D. M. Dewey, Rochester, New York, and have his imprint. (Author's collection)

By checking the imprints of the lithographers whose signatures are on the sixteen plates in the *Bloomington Nursery* collection with city directories, it is possible to chart a time span in which these plates were printed. All date from c. 1872 to c. 1876. The firms, with the approximate dates when they printed the plates are: "J. Knauber & Co. Lith. Milwaukee, Wis." (first listed under this firm name in the Milwaukee city directory for 1873–74); "R. K. Pike & Co. Lith. Louisville" (not before 1874, after about 1875); "Charles Shober, Chicago," with six subsequent changes in firm name and addresses (about 1870–76).

11. A number of these plates were examined under high magnification in March 1982 by Anne Clapp, then head of the paper conservation section of the Winterthur Museum's conservation division; she has since retired to private practice.

The black-and-white plates printed by the various lithography firms appear to have been returned to the nursery for tinting. The lines drawn in tusche create sharp, solid lines in contrast with the softer lines drawn in chalk on stone. Gum was sometimes added to the watercolors used for tinting the Phoenix plates to give luster to the fruit. Although this limited examination of the plates showed that their production was technically involved, questions and uncertainties remain. A definitive report would require detailed study.

For a description of the transfer process used since the early days of lithography see Garo Z. Antreasian and Clinton Adams, *The Tamarind Book of Lithography; Art & Techniques* (New York: Harry N. Abrams, 1971), p. 227.

12. "The Bloomington Nursery," in the Bloomington *Weekly Pantagraph*, March 20, 1870.

13. Meehan's praise appears in his *Gardener's Monthly and Horticultural Advertiser* (Philadelphia), n.s., vol. 2, no. 8 (August 1869), p. 245.

14. "The Bloomington Nursery," op. cit., n. 12.

15. "Prestele, Joseph, artist Phoenix nursery; res Walnut nw cor Evans." (*The Bloomington City Directory . . . for 1872–73* [Bloomington, Illinois: published by Edward Arntzen, 1872]), p. 268.

16. *American Agriculturist* (New York), vol. 30, no. 1 (January 1871), pp. 30, 38.

17. *Wholesale Price List . . . of the Bloomington Nursery . . . Bloomington, McLean County, Illinois. For the Spring of 1875* (University of Delaware Library); D. M. Dewey, *Dewey's Classified Catalogue of Colored Fruit Plates* (1878).

18. No place of employment was given for him in the Bloomington directories from 1876 to 1880. Foote does not appear to have been disheartened by his setback. He was an optimist; his mind was filled with schemes for fame and fortune. In May 1882, a Bloomington paper reported that "Al Foote the well known Bloomington artist, visited Missouri immediately after the slaughter of Jesse James and sketched several scenes of the outlaw's life. These he has had handsomely lithographed in colors by a Chicago house, has secured a copyright, and has brilliant prospects of realizing immensely on the sale." He also had a patent for a "contrivance to prevent accidents on cable cars." *Bloomington Daily Leader*, May 1, 1882, p. 4, col. 4. Also: *The Bloomington City Directory . . . for 1872–73*, *Gould's Bloomington and Normal Directory, for 1873*, and *Gould's Bloomington & Normal Directory for 1874–75*. Data supplied by Greg Koos, archivist, McLean County Historical Society, Bloomington, Illinois.

19. See n. 10.

20. "Prestele & Littlefield [W.H. Prestele and L. B. Littlefield], pub. colored fruit and flower plates, 401 N. Main, 2d story." *Gould's Bloomington and Normal Directory, for 1873* (Bloomington, Illinois: David B. Gould, Publisher, n.d.), p. 157.

21. The large showcard (72 × 57.6 cm., untrimmed) was printed by Charles Hamilton & Co., St. Louis, "Published by Wm. H. Prestele, Bloomington, Ills.," and copyrighted in 1872. An untrimmed copy of this print is in the Prints & Photographic Department, Library of Congress. The framed tribute to Charles Downing (about 67 × 56 cm. trimmed) is at Crawford House, the museum and headquarters of the Historical Society of Newburgh Bay and the Highlands, New York. According to Miss Helen Ver Nooy Gearn, City Historian of the City of Newburgh, the tribute was presented to the society by Mr. and Mrs. Wesley Waite (Mrs. Waite having a Downing family connection). (Miss Gearn to the author, August 9, 1981.)

22. The census of 1880 for Lucas Township, Johnson County, Iowa, gives Francis J. Prestele as being aged ten and born in Illinois; Emma C., aged eight, also born in Illinois; their mother was a native of Ireland. It has not been learned if the wife Ann died in Bloomington or after the family moved to Coralville, Iowa.

23. The account of Karolina's death is from the *Inspirations-Historie*, quoted in translation by Henrietta M. Ruff to the author, February 16, 1981.

24. U.S. Census for 1880, Lucas Township, Johnson County, Iowa, lists "Lilly," a daughter of William H. and Susanna Prestele (a native of Baden), aged two.

25. The catalogue, printed by the "Republican Job Print, Iowa City, Iowa," is 15 × 23.8 cm., with eight pages and printed wraps. (Author's collection.)

26. Karen Laughlin, librarian, Iowa State Historical Society, Iowa City, to the author, February 21, 1982.

27. The U. S. Census of 1880 for Lucas Township, Johnson County, Iowa.

28. "Announcement," in the William Henry Prestele catalogue. See n. 25.

29. William Henry Prestele is not listed in the Iowa City directories, published for 1875–76, and 1878–79, because he was living outside the city limits at Coralville and, possibly, carrying on his work there.

30. County Registrar, Vital Statistics, Johnson County, Iowa, to the author, December 18, 1981.

31. The division of pomology was created July 1, 1886. For references to Prestele's appointment see: *Official Register of the United States*, vol. 1 (Washington: Government Printing Office, 1895), p. 814; *Report of the Commissioner of Agriculture, 1887* (Washington: Government Printing Office, 1888), p. 627. Prestele's appointment shows him as a resident of Johnson County, Iowa.

32. *Report of the Commissioner of Agriculture, 1887, p. 627.* The first chief of the division was G. Onderdonk, a Texan, who had made a reputation for his research in testing and developing fruits that could be successfully grown under drought and other adverse conditions in Texas. He had established the Mission Valley Nurseries, in Mission Valley, Victoria County, Texas, sometime in the 1870s. His catalogue for 1878–79 is seventeen pages long and lists flowers, shrubs, and many varieties of fruits. (Author's collection.)

33. The quality of the chromolithography used in producing these plates varied greatly. Some of the Hoen plates were printed in their "Lithocaustic process," a forerunner of the present-day halftone. A brief description of that process and a history of the Hoen firm is in *Lithographic Lives," by Martin Wiesendanger, American-German Review*, vol. IX, no. 4(April 1943), pp. 4ff.

34. The author is indebted to J. R. McGrew, plant pathologist, USDA Fruit Laboratory, Beltsville Agricultural Research Center, Beltsville, Maryland, for the information on which my summary of the Munson-Prestele grape project is based. That story is detailed in McGrew's "The W. H. Prestele Paintings of the American Species of Grapes," *The T. V. Munson Memorial Vineyard Report*, vol. 1, no. 2 (Grayson County College, Denison, Texas, April 1, 1981.) The article includes a color reproduction of Prestele's *Vitis Lincecumii*. Sixteen of the Prestele plates, greatly reduced and in black-and-white, are in Pierre Galet's *A Practical Ampelography, Grapevine Identification*. Translated and Adapted by Lucie T. Morton; Foreword by Leon D. Adams. (Ithaca and London: Comstock Publishing Associates, a division of Cornell University Press, 1979.) Photocopies of the following were kindly supplied by Dr. Frederick G. Meyer, supervisory botanist, U.S. National Arboretum, Washington, D. C.; J. R. Magness, head horticulturist, USDA, to Philip M. Wagner, Editor, The Baltimore *Evening Sun*, May 6, 1942; "Notes on T. V. Munson's Grape Monography Manuscript. (Based on Recollections . . . of events in 1891–5 and later)", by Dr. Wm. A. Taylor, Columbus, Ohio, August 7, 1940 (Typescript); and L.H. Bailey, "The Species of Grapes Peculiar to North America," *Gentes Herbarium*, vol. III, no. 4 (Ithaca, New York, March 15, 1934), pp. 161–63. Bailey also lists the titles of the twenty-nine watercolors by Prestele. See also; James J. White and Erik A. Neumann, "The Collection of Pomological Watercolors at the U.S. National Arboretum." *Huntia*, vol. 4, no. 2 (1982). It names the artists employed by the U. S. Department of Agriculture and gives a list of the pomological plates published in the USDA *Reports* (renamed *Yearbooks* beginning in 1894).

Color Plates

Plate 1 *Strelitzia Reginae*

Strelitzia Reginae
(Queen's Bird of Paradise)
Watercolor, by Joseph Prestele.
Plate 13 in *Icones plantarum selectarum*, 1812.
Courtesy, the Bavarian State Library, Munich, West Germany, and reproduced here by permission.

Plates 1–3

The portfolio, *Icones plantarum selectarum Horti Regii Academii Monacensis*, by Joseph Prestele, 1812, of forty-eight watercolors of plants, was produced entirely by the sixteen-year-old artist.

Strelitzia
Reginae.
13.

Plate 2 *Passiflora Alata*

Passiflora Alata
(Wingstem Passiflora)
Watercolor, by Joseph Prestele.
Plate 19 in *Icones plantarum selectarum*, 1812.
Courtesy, the Bavarian State Library, Munich, West Germany,
and reproduced here by permission.

Passiflora alata.
19.

Plate 3 *Ferraria Pavonia*

Ferraria Pavonia. Pfauenartige Ferrarie
(*Tigridia pavonia.* Common Tigerflower)
"Joseph Prestele pinx."
Watercolor.
Plate 32 in *Icones plantarum selectarum,* 1812.
Courtesy, the Bavarian State Library, Munich, West Germany,
and reproduced here by permission.

Ferraria Pavonia
Pfauenartige Ferrarie
32.

Plate 4 *Prunus Tomentosa*

Prunus Tomentosa
"de Villeneuve del. J. Prestele sc."
Plate 23 in Philipp Franz von Siebold, *Flora Japonica*, 2 vols. (Leipzig, 1835–44), vol. 2.

Prestele engraved the plate from a drawing by de Villeneuve, a botanical artist (of Munich?).

Plates 4–7

Joseph Prestele's European career as a botanical artist began in Bavaria about 1816 when he was twenty, and continued with increasing recognition among botanical authors and publishers until 1843 when he and his family, along with other co-religionists, emigrated to America.

During those years he was employed first as a staff artist at the Royal Botanical Garden in Munich for almost a decade. While there, he met and became friends with many leading botanists, some of whom gave him his first commissions. After 1825, when he was self-employed, he collaborated in illustrating some of the greatest European botanical works of the time. Unfortunately, much of his work is unsigned.

Each commission made different demands on Prestele's many skills as an artist. Sometimes he made the original drawings for the plates; sometimes it was necessary to revise the drawings made by botanists who were clumsy with a pencil. Much of his work involved making scientifically accurate and visually appealing engravings on stone whose lines were drawn with infinite delicacy so that the hand coloring would appear especially brilliant. Occasionally he performed all the steps in the production of an illustration (except the printing), beginning with the original drawing, making the engraving, and finally coloring the printed plate.

By the time that Joseph Prestele left for America in 1843, he had become known in botanical circles as one of the few artists skilled in many different specialties. Success, however, brought little financial reward to this talented man. The field of botanical art was a limited one; its practitioners, even the best ones, were poorly paid. Prestele's life was one of poverty in which he was sustained by his patient and self-effacing wife, his love of nature, his art, and his abiding religious faith.

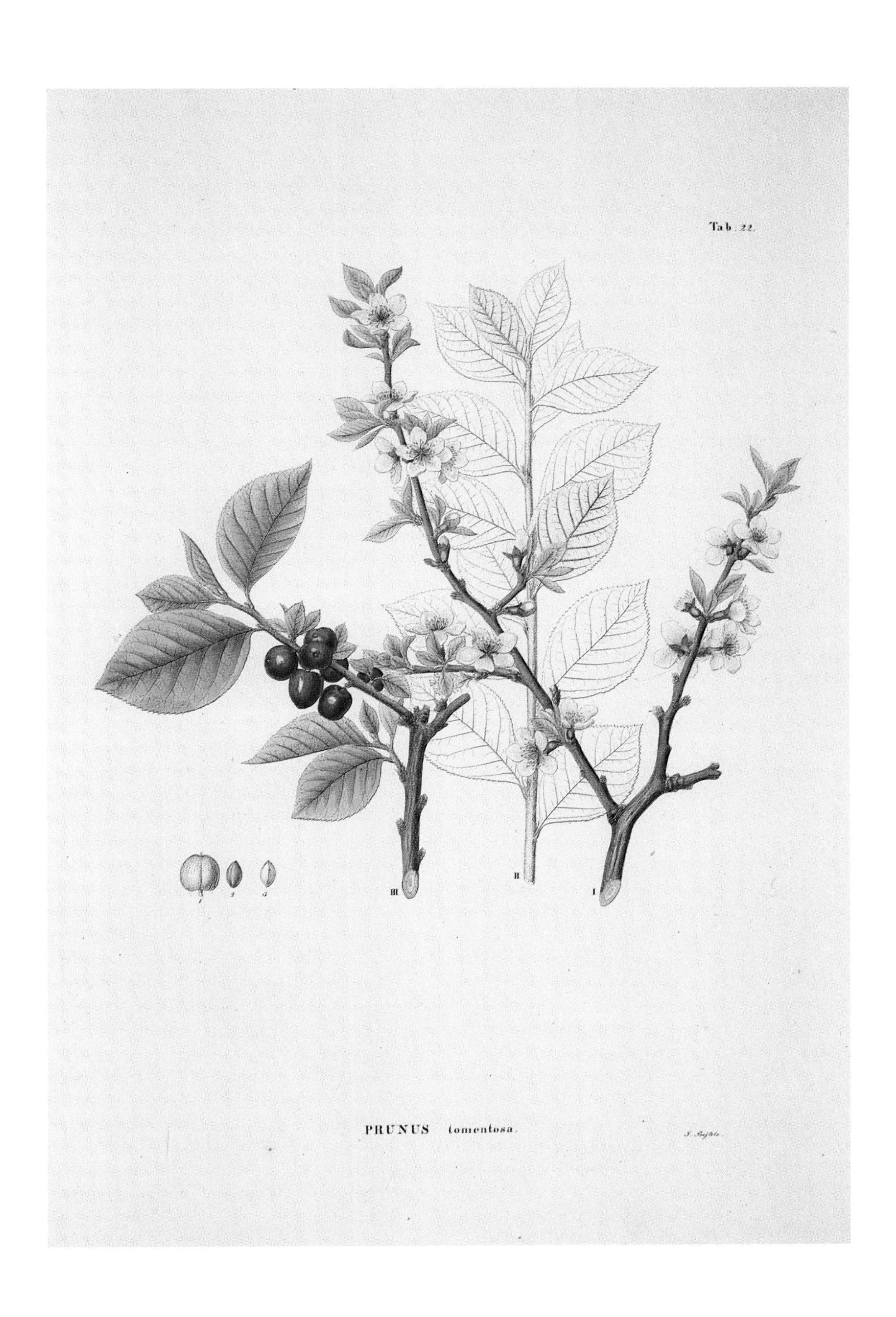
Tab: 22.
I
II
III
1
2
3
PRUNUS tomentosa.

Plate 5 *Benthamia Japonica*

Benthamia Japonica
(*Cornus Kousa*)
"Popp del. J. Prestele sc."
Lithograph; engraved on stone, colored.
Plate 16 in Philipp Franz von Siebold, *Flora Japonica*, 2 vols. (Leipzig, 1835–44), vol. 2.

H. Popp (1830–1870), who made the drawing, was a botanical artist of Bamberg, West Germany.

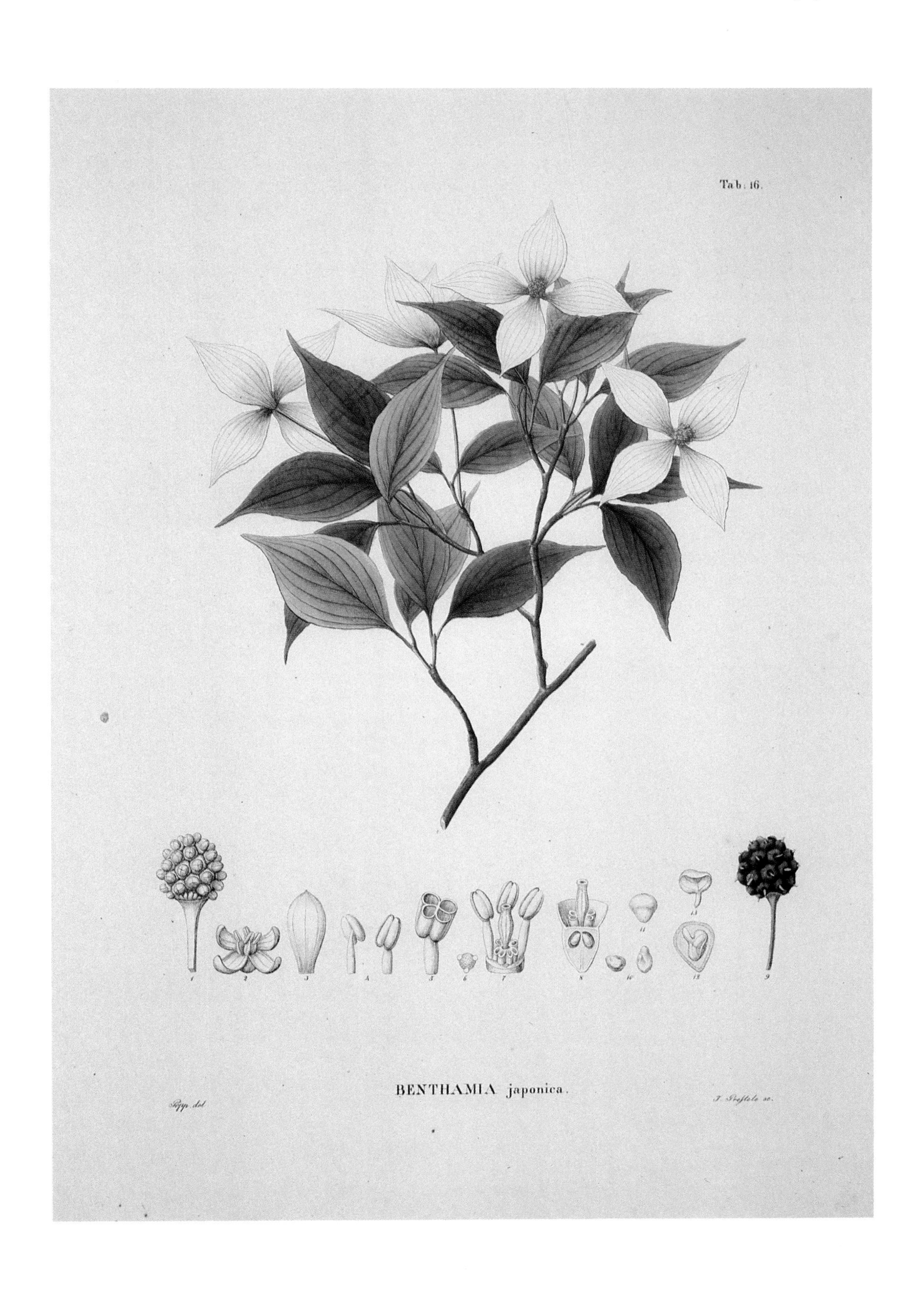
Tab: 16.
BENTHAMIA japonica.

Plate 6 *Picea Obovata*

Picea Obovata
(*Picea obovata.* Siberian spruce)
Lithograph; engraved on stone, colored; 32 × 19 cm.
Plate 499 in Karl F. von Ledebour, *Icones plantarum novarum vel imperfecte cognitarum, floram rossicam.*
5 vols. in 3. (Riga, etc., 1829–34), vol. 5.
Courtesy, the Arnold Arboretum of Harvard University.

Tab. 499.
Picea obovata

Plate 7 *Prenanthes Polymorpha*

Prenanthes Polymorpha
(*Prenanthes --?--*)
"E. Bremmer ad. sicc. del. Prestele sc."
Lithograph; engraved on stone, colored; 32 × 19 cm.
Plate 498 in Karl F. von Ledebour, *Icones plantarum novarum vel imperfecte cognitarum, floram rossicam.*
5 vols. in 3. (Riga, etc., 1829–34), vol. 5.
Courtesy, the Arnold Arboretum of Harvard University.

Tab. 498
Prenanthes polymorpha
a
b

Plate 8 *Conium Maculatum*

Gefleckter Schierling. Conium Maculatum
(Spotted hemlock)
Unsigned lithograph; drawn, engraved on stone, and colored by Joseph Prestele; 43 × 27 cm.
Plate 8 from *Die wichtigsten Giftpflanzen Deutschlands* (Friedberg, Hessen: Carl Bindernagel, 1843).
Courtesy, Gray Herbarium of Harvard University.

Plates 8–11

In 1837 Joseph Prestele and his family moved with other Inspirationists to the buildings of a former convent at Engelthal, near Friedberg, Hessen, which the Community of True Inspiration had leased. There the artist met Carl Soldan, who had studied theology—presumably with the intention of becoming a minister, but because of illness, had instead become a teacher at the seminary in Friedberg. Soldan was an amateur artist and botanist and was particularly interested in the local flora, which he studied on weekly field trips with his students. He became concerned by the general lack of knowledge about the dangers of the poisonous native plants. From that concern, and because of the mutual interests in botany and art which he and Prestele shared, came a beautiful publication *Die wichtigsten Giftpflanzen Deutschlands*, published in 1843 in folio size; Soldan wrote the fifty-six pages of text, and Prestele drew the twenty-four plates illustrating the plants life-size. Their friend Carl Bindernagel was the publisher, and he also printed the work at his small shop in Friedberg.

The long descriptive title explains the purpose of the book and the contributions made by each of the three men to it. The title can be translated as: The most important poisonous plants of Germany in life size illustration, as warning and instruction about their danger, painted true to nature and engraved in stone by Joseph Prestele, flower painter and lithographer, selected and described by Carl Soldan, second teacher of the evangelical teacher-seminary in Friedberg. (Friedberg in the Wetterau. Printed and published by Carl Bindernagel, 1843.)

Four plates from this very scarce and beautiful work are reproduced here. They are from a copy in the library of the Gray Herbarium of Harvard University. The copy is apparently the one that Joseph Prestele gave to Asa Gray in September 1843.

Nº 8.
a
b
c
d
Conium maculatum
Carl Bindernagel

Plate 9 *Aconitum Störkianum*

Blauer Sturmhut. Aconitum Störkianum
(*Aconitum bicolor*, Blue monkshood)
Unsigned lithograph; drawn, engraved on stone,
and colored by Joseph Prestele; 43 × 27 cm.
Plate 12 from *Die wichtigsten Giftpflanzen Deutschlands*
(Friedberg, Hessen: Carl Bindernagel, 1843).
Courtesy, Gray Herbarium of Harvard University.

№ 12
Blauer Sturmhut. Aconitum-Störkianum.
Druck u. Verlag von C. Bindernagel in Friedberg i. d. Wetterau.

Plate 10 *Clematis Vitalba*

Gemeine Waldrebe. Clematis Vitalba
(Traveler's Joy)
Unsigned lithograph; drawn, engraved on stone,
and colored by Joseph Prestele; 43 × 27 cm.
Plate 17 from *Die wichtigsten Giftpflanzen Deutschlands*
(Friedberg, Hessen: Carl Bindernagel, 1843).
Courtesy, Gray Herbarium of Harvard University.

No 17
a
b
c
Clematis Vitalba.

Plate 11 *Colchicum Autumnale*

Herbst-Zeitlose. Colchicum Autumnale
(Common autumn crocus, Meadow saffron)
Unsigned lithograph; drawn, engraved on stone,
and colored by Joseph Prestele; 43 × 27 cm.
Plate 20 from *Die wichtigsten Giftpflanzen Deutschlands*
(Friedberg, Hessen: Carl Bindernagel, 1843).
Courtesy, Gray Herbarium of Harvard University.

№ 20.
a
c
b f e d
Colchicum autumnale.
Druck u. Verlag von C. Bindernagel in Friedberg i. d. Wetterau.

Plate 12 *Brazoria Truncata*

Brazoria Truncata
(*Brazoria truncata*. Rattlesnake flower)
"Sprague omnes del. Prestele omnes in lap. sc."
"Hall & Mooney exc.," Buffalo, New York.
Lithograph; engraved on stone, colored; 28.5 × 23 cm.
Plate 5 in Asa Gray, "Chloris Boreali-Americana," 1846.
Courtesy, Botany Library, Smithsonian Institution.

Plates 12–17

In January 1845, when Joseph Prestele wrote introducing himself to Asa Gray and asking for commissions, Gray—then professor of botany at Harvard University—was planning a new work on North American plants. He needed someone to make engravings from drawings by a New England artist, Isaac Sprague, and to color the printed plates. Prestele's credentials delighted Gray and he gave the artist the commission. The following year Gray's article appeared with the title "Chloris Boreali-Americana. Illustrations of New, Rare, or Otherwise Interesting North American Plants, Selected Chiefly from Those Recently Brought into Cultivation at the Botanic Garden of Harvard Univeristy. Part I." It appeared in the *Memoirs* of the American Academy of Arts and Sciences, Volume 3, with fifty-six pages of text and ten plates engraved and colored by Prestele, from drawings by Sprague, and printed by Hall & Mooney, Buffalo, New York. Gray had intended to publish further numbers but instead he dropped that plan and began working on another publication. Prestele's contribution to the "Chloris" was his introduction to the field of botanical illustrating in America, and the beginning of his association with the small number of American botanists then engaged in scientific research and publishing.

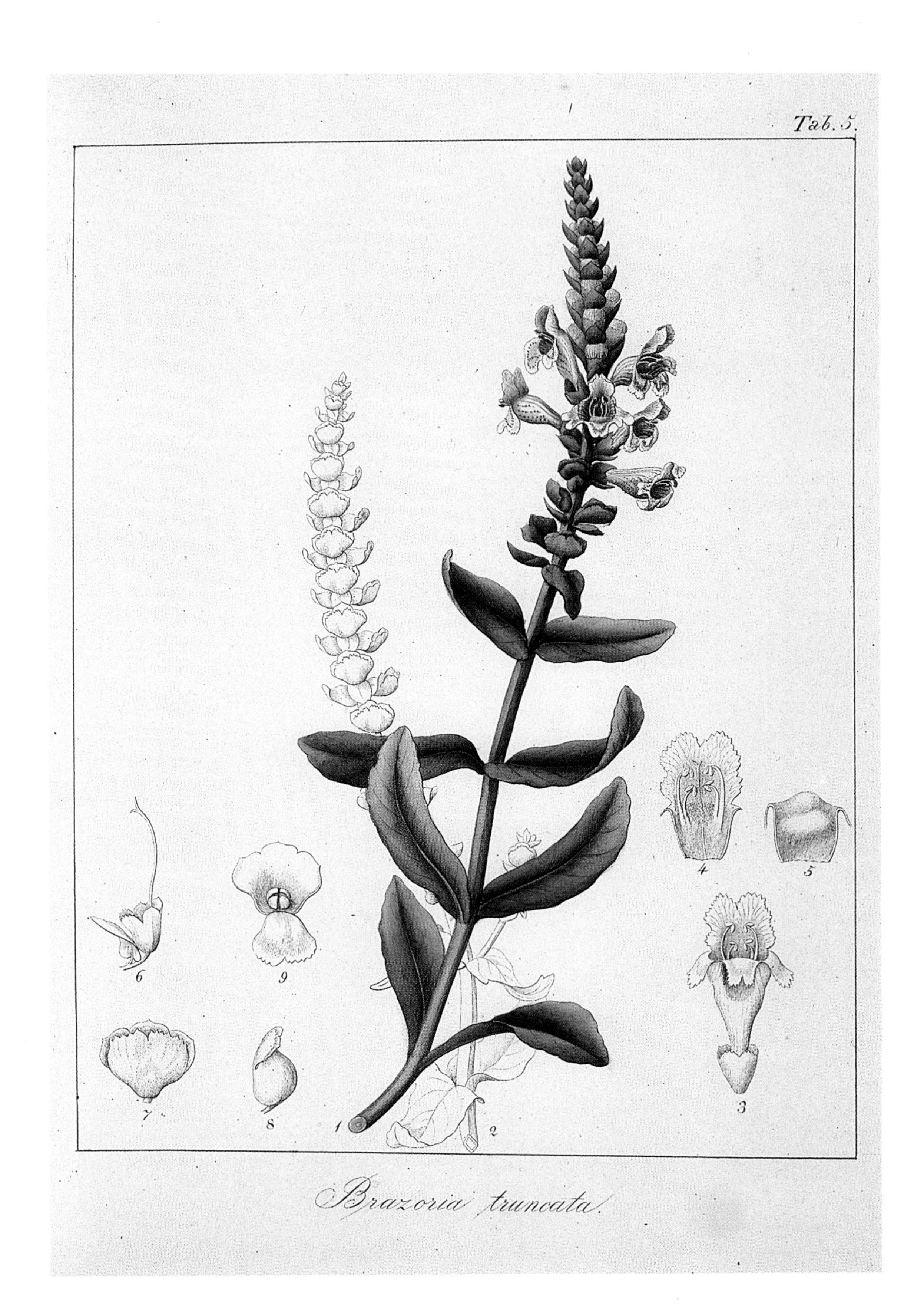
Tab. 5.
1
2
3
4
5
6
7
8
9
Brazoria truncata.

Plate 13 *Gaillardia Amblyoden*

Gaillardia Amblyoden
(*Gaillardia amblyodon*. Maroon Gaillardia)
"Sprague omnes del. Prestele omnes in lap. sc."
"Hall & Mooney exc.," Buffalo, New York.
Lithograph; engraved on stone, colored; 28.5 × 23 cm.
Plate 4 in Asa Gray, "Chloris Boreali-Americana," 1846.
Courtesy, Botany Library, Smithsonian Institution.

Prestele had many problems with this plate, as he wrote Gray in February 1846. First, he had found that Hall & Mooney were inexperienced in printing such engravings and their impressions came off too pale. Then he tried to correct the Gaillardia engraving according to Gray's instructions, but that led to a succession of disasters. He finally became so nervous and confused that he made mistakes on other plates and had to rework them. "In short, Dear Sir, I suffered a great deal on account of seeing that I was unable to do Things Well for you."

Tab. 4.
3
4
5
1
2
Gaillardia amblyodon.

Plate 14 *Oakesia Conradii*

Oakesia Conradii
(*Uvularia* [*grandiflora?*]. Bellwort, Merrybells)
"Sprague omnes del. Prestele omnes in lap. sc."
"Hall & Mooney exc.," Buffalo, New York.
Lithograph; engraved on stone, colored; 28.5 × 23 cm.
Plate 1 in Asa Gray, "Chloris Boreali-Americana," 1846.
Courtesy, Botany Library, Smithsonian Institution.

Tab. 1.
1
2
3
4
5
6
7
8
9
10
11
12
13
14
15
Sprague omnes del.
Prestele omnes in lap. fc.
Hall & Mooney exc.
Oakesia Conradii.

Plate 15 *Thermopsis Caroliniana*

Thermopsis Caroliniana
(*Thermopsis caroliniana*. Carolina thermopsis, Bush-pea)
"Sprague omnes del. Prestele omnes in lap. sc."
"Hall & Mooney exc.," Buffalo, New York.
Lithograph; engraved on stone, colored; 28.5 × 23 cm.
Plate 7 in Asa Gray, "Chloris Boreali-Americana," 1846.
Courtesy, Botany Library, Smithsonian Institution.

Tab. 7.
2
1
Thermopsis caroliniana.

Plate 16 *Thermopsis Fraxinifolia*

Thermopsis Fraxinifolia
(*Thermopsis fraxinifolia.* Ashleaf T.)
"Sprague omnes del. Prestele omnes in lap. sc."
"Hall & Mooney exc.," Buffalo, New York.
Lithograph; engraved on stone, colored; 28.5 × 23 cm.
Plate 8 in Asa Gray, "Chloris Boreali-Americana," 1846.
Courtesy, Botany Library, Smithsonian Institution.

Tab. 8.
Thermopsis fraxinifolia.

Plate 17 *Thermopsis Mollis*

Thermopsis Mollis
(Soft Thermopsis)
No signatures; from a drawing by Isaac Sprague;
the plate engraved and colored by Joseph Prestele;
printed by Hall & Mooney, Buffalo, New York; 28.5 × 23 cm.
Plate 9 in Asa Gray, "Chloris Boreali-Americana," 1846.
Courtesy, Botany Library, Smithsonian Institution.

Tab. 9.
Thermopsis mollis.

Plate 18 *Tyson Pear*

Tyson Pear
"Lithd. of J. Prestele"
Lithograph; engraved on stone, colored; 26 × 17.5 cm.
In *Transactions* of the Massachusetts Horticultural Society, vol. 1, no. 2 (1848).
Courtesy, New York Botanical Garden.

Joseph Prestele explained to Asa Gray in 1849 that by engraving plates with very delicate lines, "I can execute the coloring very well." He used as examples the eight plates he made for the Massachusetts Horticultural Society's *Transactions*.

The design of this plate is unlike two other versions that also appear to be the work of Joseph Prestele. The earliest of these is "The Tyson Pear," lithographed by E. B. & E. C. Kellogg, Hartford, Connecticut, with the legend, "Kelloggs Series of Fruits, Flowers and Ornamental Trees. 28." (28.5 × 22.5 cm.) It is included in D. M. Dewey, *The Colored Fruit Book for the Use of Nurserymen* (Rochester, N.Y., 1859). It is engraved on stone and is hand colored. A comparison of this plate made for the *Transactions* with one of the Tyson pear published in the *Horticulturist, A Journal of Rural Art and Rural Taste* (vol. 10, no. 2, February 1855, opposite p. 57), reveals that the *Horticulturist* plate is an edited version of the Kellogg print, with elements removed to adapt the design to the smaller size of the magazine page. This evidence is particularly interesting because it proves that the Kelloggs were supplying plates for the nursery trade before Dewey entered the field (ca. 1858), and that they may have commissioned Prestele to make some of the engravings for them.

The Tyson pear, which ripens in late August and early September, originated as an accidental seedling in a hedge row on the farm of Jonathan Tyson of Jenkintown, near Philadelphia, during the 1790s. For many years it aroused little more than local interest but when examples of its fruit attracted favorable attention at a meeting of the American Pomological Society, nurserymen and growers became interested. By 1855 the pear began to be disseminated and its sale promoted.

Plates 18–19

In 1847 the Committee of Publication of the Massachusetts Horticultural Society in Boston, then in the full tide of its success, asked Joseph Prestele if he would make four colored engravings for the second number of the society's new *Transactions*. The committee must have stressed that the plates were to be produced in a "high standard of excellence," for they had just had an unfortunate experience with their first effort.

Some years before, an English lithographer, William Sharp, had introduced chromolithography to Boston. The process was then relatively new and not fully developed, but it had

[*Continued on page 166*]

TYSON PEAR.

Plate 19 *Red Astrachan Apple*

Red Astrachan Apple
Unsigned lithograph; engraved on stone and colored by Joseph Prestele; 26 × 17.5 cm.
In the *Transactions* of the Massachusetts Horticultural Society, vol. 1, no. 3 (1849).
Courtesy, New York Botanical Garden.

Andrew Jackson Downing spoke of the Red Astrachan as an apple of extraordinary beauty whose rich color was heightened by an exquisite bloom on the surface of the fruit, like that of a plum. He thought it one of the handsomest dessert fruits and of good quality if picked as soon as it ripened, from mid-July to mid-August.

This apple, of Russian origin, is said to have been brought to England from Sweden in 1816. Examples of the fruit were exhibited at the Massachusetts Horticultural Society in 1835, and it began to be grown in America at that time. Growers found that the tree was vigorous and hardy. During its introduction to various parts of America it accumulated a variety of names including "Early Rus," "Red Ashmore," "Vermillion d'Ete," "Astrachaner rother," and, unaccountably, "Abe Lincoln."

[*Continued from page 164*]

great possibilities for mechanically producing colored pictures. Members of the Committee of Publication liked the samples shown them, and they were pleased with Sharp's sketches of the proposed designs. The finished plates, however, were a bitter disappointment, largely because of the unsatisfactory coloring. After considerable confusion, the committee finally released the plates but with the promise of giving up "chromolithing," and returning to the beautiful handwork long used in the finest horticultural and botanical works. It was at that point that they approached Prestele. He accepted the commission and provided four plates for each of the succeeding two numbers.

The Sharp incident was, of course, more than a matter of unsatisfactory work. Instead, it was a confrontation of sorts between old ways and new; between the expensive hand artistry that was giving way to processes which the age of mechanization was bringing in. With the Massachusetts Horticultural Society and Prestele, the old way of producing botanical illustrations had only a brief respite. A few years after that victory, Louis Prang, a young German from Breslau, came to Boston. Within a generation he had shown America the delicate and realistic capabilities of a fully developed chromolithography.

RED ASTRACHAN APPLE.

Plate 20 *American Crab Apple*

Pyrus Coronaria. American Crab Apple
(*Malus coronaria* L. Mill. Sweet crabapple, crabapple, wild crab, garland-tree)
"I. Sprague del. J. Prestele sc."
Lithograph; engraved on stone, printed, and colored by Joseph Prestele; 34 × 27 cm.
Plate 52 for the projected *Forest Trees of North America*, by Asa Gray.

Plates 20–38

In 1846, when Joseph Henry (1797–1878) became the first secretary of the Smithsonian Institution, he had the responsibility of organizing the new institution and defining its scope and direction. James Smithson's will had stipulated that his bequest should be used for the "increase and diffusion of knowledge among men," a directive with which Henry was in complete agreement. His background was that of a teacher, a physicist, and a mathematician. His interests were far ranging. He believed that the Smithsonian's program should be designed to interest and serve the general public as well as those in advanced fields of knowledge.

Henry's first plans included a series of books, Reports on the Progress of Knowledge. He decided that the first report should be a well-illustrated study of North American trees, which would include both scientific descriptions and their economic uses. Asa Gray, then professor of botany at Harvard, agreed to direct the project and write the text; the Smithsonian would cover the expenses.

Gray engaged two artists with whom he had worked before to start on the illustrations; the young New Englander Isaac Sprague, who was assigned to make the original drawings, and Joseph Prestele, a German immigrant, who was to cut the engravings, color the plates, and, as events turned out, do some of the printing. Technical and production problems developed as the work progressed and Gray made an exacting and sometimes irritating taskmaster.

By the early 1850s some twenty-three plates had been completed; apparently these particular subjects were done first because specimens of buds, leaves, flowers, and seeds throughout the growing season were readily available. Then, for a variety of reasons, the project was first stalled and then finally dropped.

The plates, so painstakingly created, were put away and forgotten until 1891 when Dr. Samuel P. Langley, secretary of the Smithsonian, distributed sets of them to various scholars and libraries both in America and abroad.

The following plates from the *Forest Trees of North America* are courtesy of the Botany Library, Smithsonian Institution.

PYRUS CORONARIA. American Crab Apple.

Plate 21 *Common Locust Tree*

Robinia Pseudacacia. Common Locust Tree
(*Robinia pseudoacacia* L. Black Locust, Yellow Locust, Locust)
"I. Sprague del. I. Prestele sc."
Lithograph; engraved on stone; printed, and colored by Joseph Prestele; 34 × 27 cm.
Plate 34 for the projected *Forest Trees of North America*, by Asa Gray.

On February 16, 1850, after Joseph Prestele had bought a lithographic press and hired a printer, he wrote Gray that he would like to print all the plates which he was to color. If Gray approved, Prestele asked that paper be sent him so he could begin making the plates of the "Robinia," a proof of which he had sent Gray. (He did not state to which of the three Robinia plates produced for the tree book he referred. From evidence in other letters it seems possible that it was this one.)

Pl. 34.
ROBINIA PSEUDACACIA. Common Locust Tree.
I. Sprague del.
I. Prestele sc.

Plate 22 *Choke Cherry*

Cerasus Virginiana. Choke Cherry
(*Prunus virginiana.* Common Chokecherry)
Unsigned lithograph; drawn by Isaac Sprague;
engraved on stone, printed, and colored by Joseph Prestele; 34 × 27 cm.
Plate 49 for the projected *Forest Trees of North America*, by Asa Gray.

Prestele sent a colored proof of the "Cerasus" plate (possibly this subject) to Asa Gray on December 19, 1849, asking Gray to let him color the final copies. "I would reduce the prize [price] to $8 per hundred," he wrote. Two months later (February 16, 1850) he again wrote Gray, this time to ask if he could both color *and* print the plates, as he had brought a press and hired a printer. Prestele added that if Gray approved, he should forward a supply of paper and write out the caption "with the exact pattern of the letters."

49.
1
2
CERASUS VIRGINIANA. Choke Cherry.

Plate 23 *Wild Black Cherry*

Cerasus Serotina. *Wild Black Cherry*
(*Prunus serotina var. serotina.* Black cherry, rum cherry, etc.)
"Sprague del. Prestele in lap. sc. et lith."
Lithograph; engraved on stone and colored by Joseph Prestele; 34 × 27 cm.
Plate 50 for the projected *Forest Trees of North America*, by Asa Gray.

CERASUS SEROTINA. Wild Black Cherry.

Plate 24 *Wild Red Cherry*

Cerasus Pennsylvanica. *Wild Red Cherry*
(*Prunus pensylvanica.* Pin Cherry, bird cherry, fire cherry, wild red cherry, etc.)
Unsigned lithograph; drawn by Isaac Sprague;
engraved on stone, printed, and colored by Joseph Prestele; 34 × 27 cm.
Plate 48 for the projected *Forest Trees of North America*, by Asa Gray.

Pl. 48.
CERASUS PENNSYLVANICA. Wild Red Cherry.

Plate 25 *Chickasaw Plum*

Prunus Chicasa. *Chickasaw Plum*
(*P. Angustifolia.* Chickasaw Plum)
"Sprague del. Prestele in lap. sc. et lith."
Lithograph; engraved on stone, and colored by Joseph Prestele; 34 × 27 cm.
Plate 47 for the projected *Forest Trees of North America*, by Asa Gray.

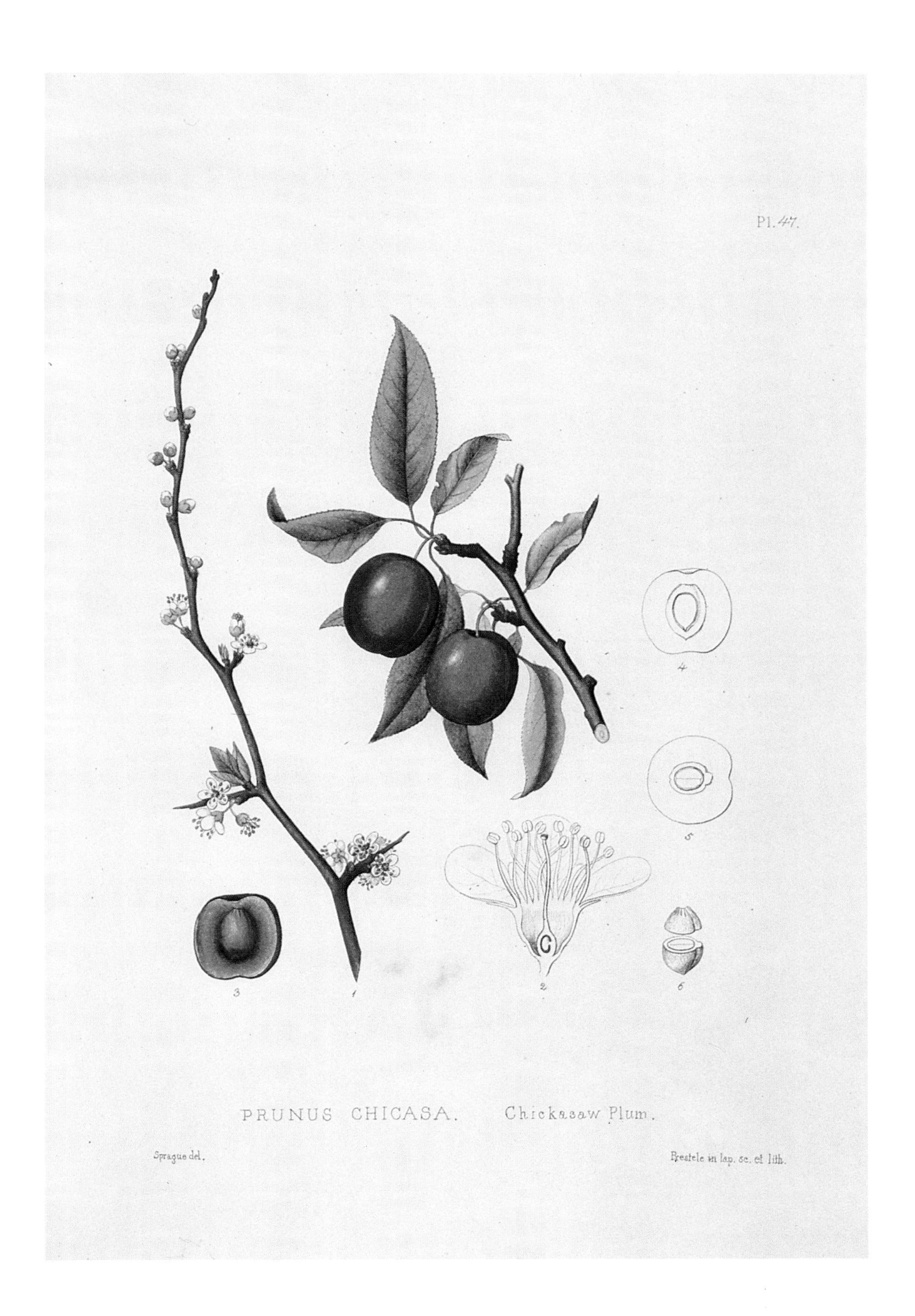
Pl. 47.
4
5
3
1
2
6
PRUNUS CHICASA. Chickasaw Plum.
Sprague del.
Prestele in lap. sc. et lith.

Plate 26 *Wild Plum*

Prunus Americana. Wild Plum
(*Prunus americana var. americana.* American plum, red plum, yellow plum, river plum, wild plum, etc.)
Unsigned lithograph; drawn by Isaac Sprague;
engraved on stone, printed, and colored by Joseph Prestele; 34 × 27 cm.
Plate 46 for the projected *Forest Trees of North America*, by Asa Gray.

Prestele informed Gray on February 11, 1851, that he had increased his charge for coloring this plate to 10 cents, which "enables me to do them so much better and with more care."

PRUNUS AMERICANA. Wild Plum.

Plate 27 *Honey Locust*

Gleditschia Triacanthos. *Honey Locust*
"Sprague del." (in script)
Lithograph; engraved on stone; printed, and colored by Joseph Prestele; 34 × 27 cm.
Plate 41 for the projected *Forest Trees of North America*, by Asa Gray.

Joseph Prestele had difficulty producing a plate of the honey locust that Gray approved. Sometime before April 1, 1852, he had printed and colored 100 plates which Gray rejected. The artist then offered to make a new engraving, without charge, and print new plates if Gray would supply paper for them, and pay $8 for the coloring. A few weeks later Gray sent the money and the paper, and the corrected proof. "The little alterations which you marked on it," Prestele wrote him, "I have done to your wishes, and will take pains to do all to your full satisfaction, especially the coloring." He reported the new paper satisfactory but thinner than an earlier supply.

Pl. 41.
GLEDITSCHIA TRIACANTHOS. Honey Locust.

Plate 28 *Kentucky Coffee Tree*

Gymnocladus Canadensis. *Kentucky Coffee Tree*
(*Gymnocladus dioicus.* Kentucky Coffee tree)
"J. Prestele sc. I. Sprague del."
Lithograph; engraved on stone, colored; 34 × 27 cm.
Plate 40 for the projected *Forest Trees of North America*, by Asa Gray.

"I engraved the plate 40 again by your advise [advice] and it stands correct now. My charge will be only $2 although it was very difficult to engrave, particularly the fruit." (Joseph Prestele to Asa Gray, February 11, 1851.)

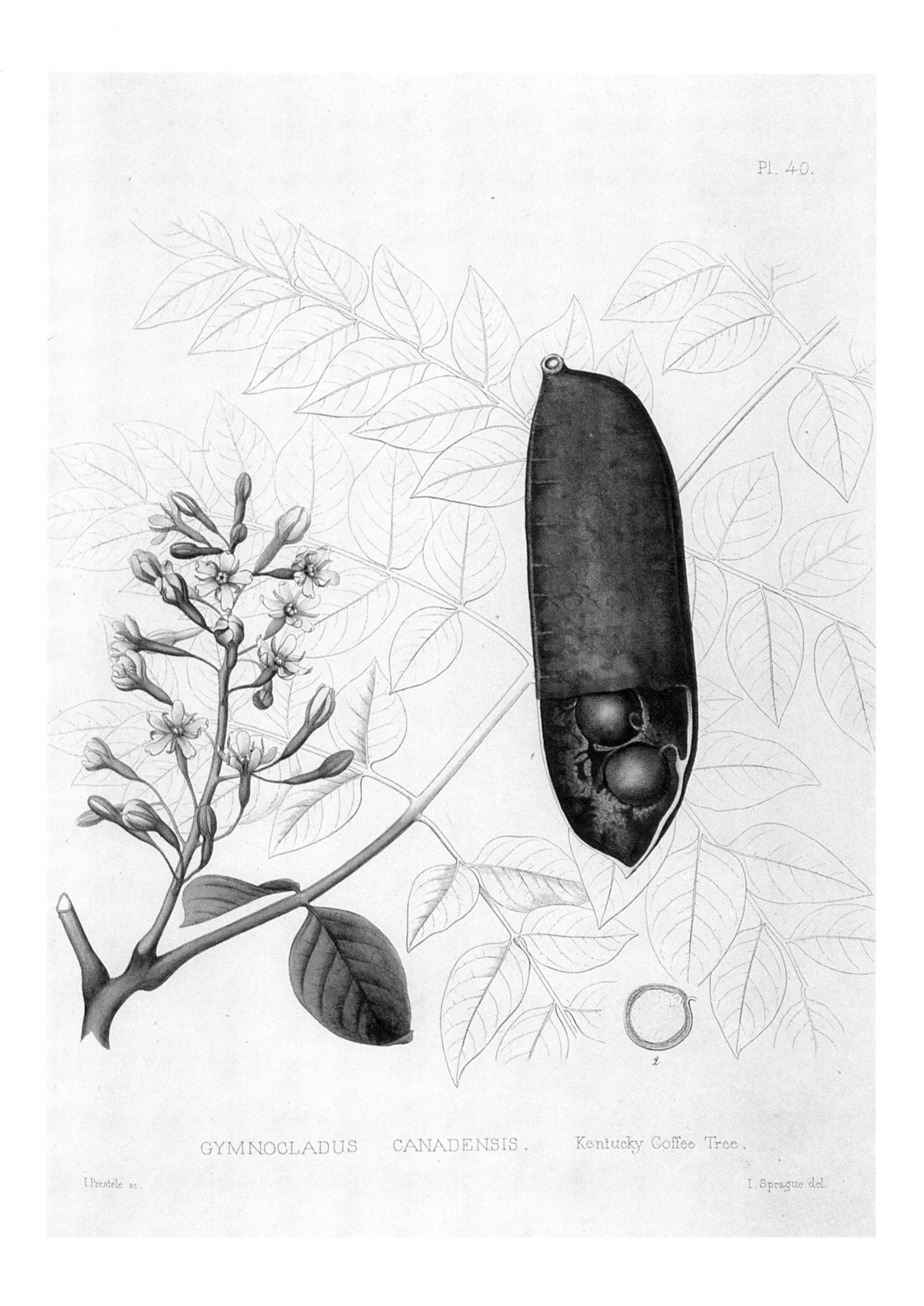
Pl. 40.
2
GYMNOCLADUS CANADENSIS. Kentucky Coffee Tree.
I. Prestele sc.
I. Sprague del.

Plate 29 *Red Bud*

Cercis Canadensis. *Red Bud*
(*Cercis canadensis var. canadensis.* Eastern redbud)
Unsigned lithograph; drawn by Isaac Sprague;
engraved on stone, printed, and colored by Joseph Prestele; 34 × 27 cm.
Plate 39 for the projected *Forest Trees of North America*, by Asa Gray.

On February 11, 1851, Joseph Prestele wrote Asa Gray that he had increased his charge for coloring this plate to 10 cents each.

CERCIS CANADENSIS. Red Bud.

Plate 30 *Clammy Locust Tree*

Robinia Viscosa. *Clammy Locust Tree*
(*Robinia viscosa var. viscosa.* Clammy Locust)
Unsigned lithograph; drawn by Isaac Sprague;
engraved on stone, printed, and colored by Joseph Prestele; 34 × 27 cm.
Plate 35 for the projected *Forest Trees of North America*, by Asa Gray.

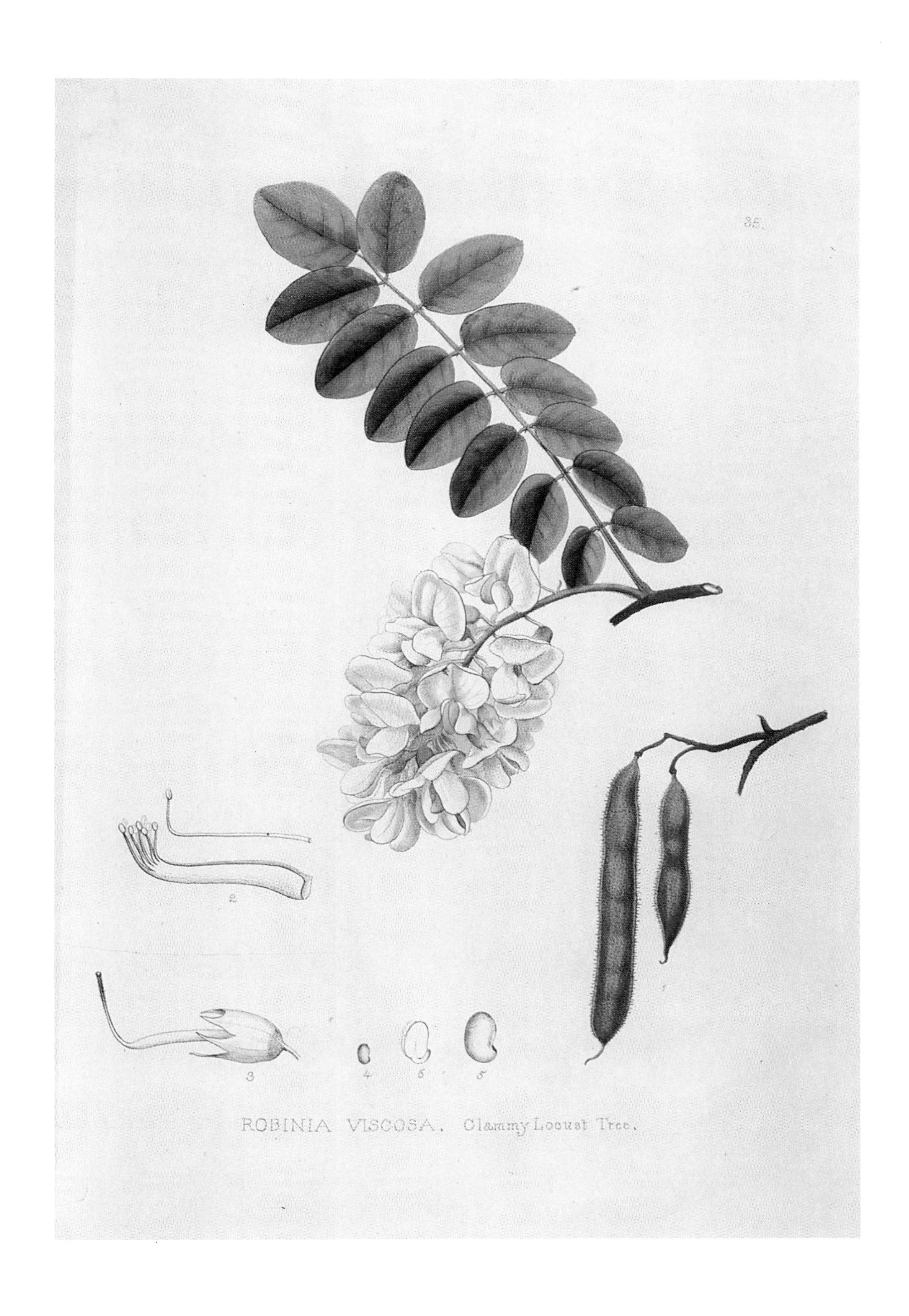

ROBINIA VISCOSA. Clammy Locust Tree.

Plate 31 *Bottlebrush Buckeye*

Aesculus Parviflora. [*Bottlebrush Buckeye*]
(*A. parviflora.* Bottlebrush Buckeye)
"Sprague del. Prestele sc. et lith."
Lithograph; engraved on stone, and colored by Joseph Prestele; 34 × 27 cm.
Plate 31 for the projected *Forest Trees of North America*, by Asa Gray.

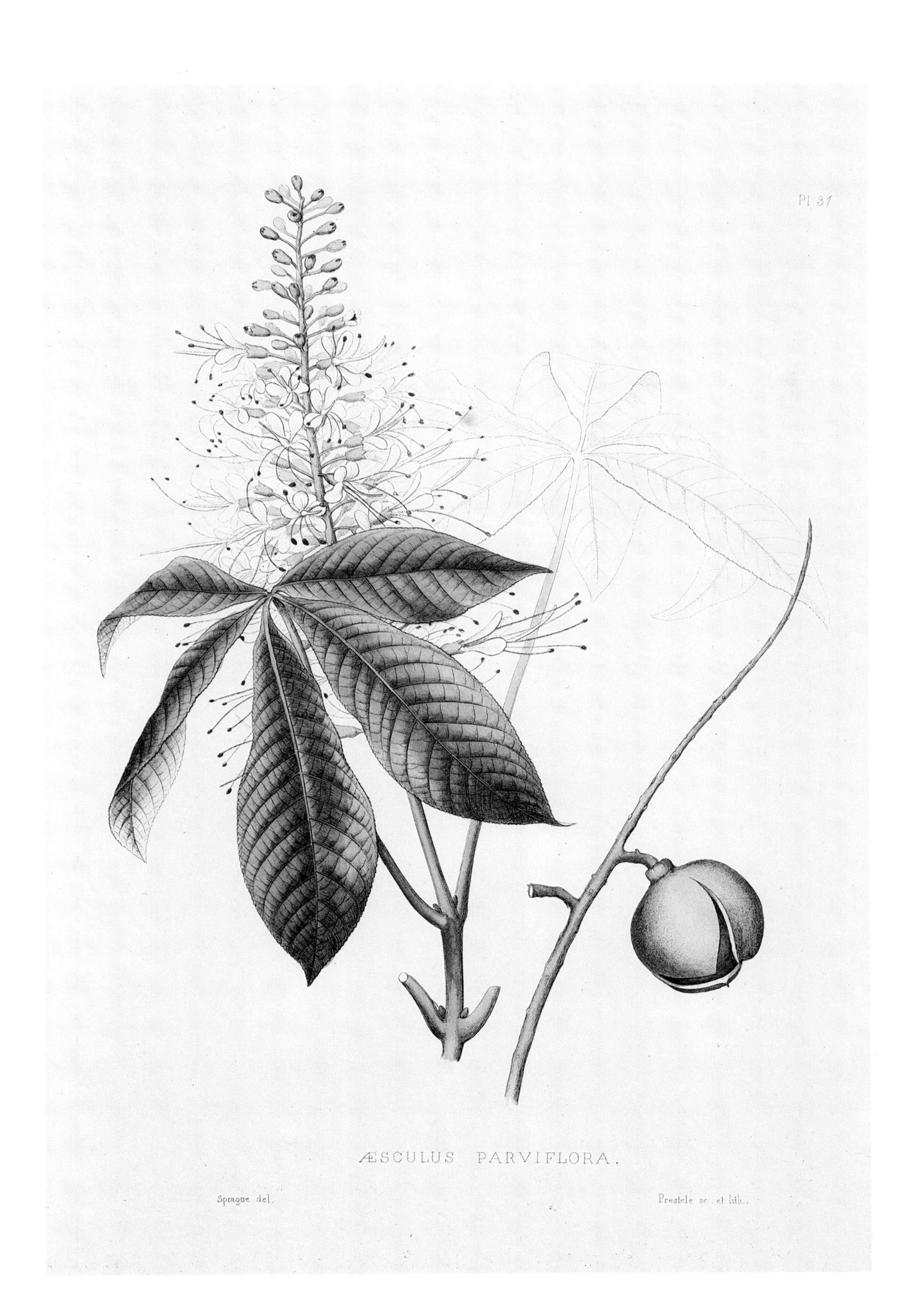
Pl. 37
ÆSCULUS PARVIFLORA.
Sprague del.
Prestele sc. et lith.

Plate 32 *Red Buckeye*

Aesculus Discolor. [*Red Buckeye*]
(*Aesculus pavia* L. Red buckeye, scarlet buckeye, woolly buckeye, firecracker-plant)
"Sprague del. Prestele sc. et lith."
Lithograph; engraved on stone, and colored by Joseph Prestele; 34 × 27 cm.
Plate 30 for the projected *Forest Trees of North America*, by Asa Gray.

Joseph Prestele increased his charge for coloring this plate to 12 cents, as he wrote Asa Gray on February 11, 1851, which "enables me to do them so much better."

Pl.30.
ÆSCULUS DISCOLOR.
Sprague del.
Prestele sc. et lith.

Plate 33 *Ohio Buckeye*

Aesculus Glabra. *Ohio Buckeye*
Unsigned lithograph; drawn by Isaac Sprague;
engraved on stone, printed, and colored by Joseph Prestele; 34 × 27 cm.
Plate 27 for the projected *Forest Trees of North America*, by Asa Gray.

Early in January 1852, Dr. Spencer Fullerton Baird, assistant secretary of the Smithsonian Institution, appears to have questioned either Prestele's charges or the amount of time he spent on producing the plates. In his reply the artist said that "the coloring of such a plate as Aesculus glabra needs indeed a little more than a common colorist can bestow or do, and since I take pain[s] to do the work myself, my wages a day is very little, because I can do very few a day." (Joseph Prestele to S. F. Baird, January 27, 1852. Smithsonian Institution Archives.)

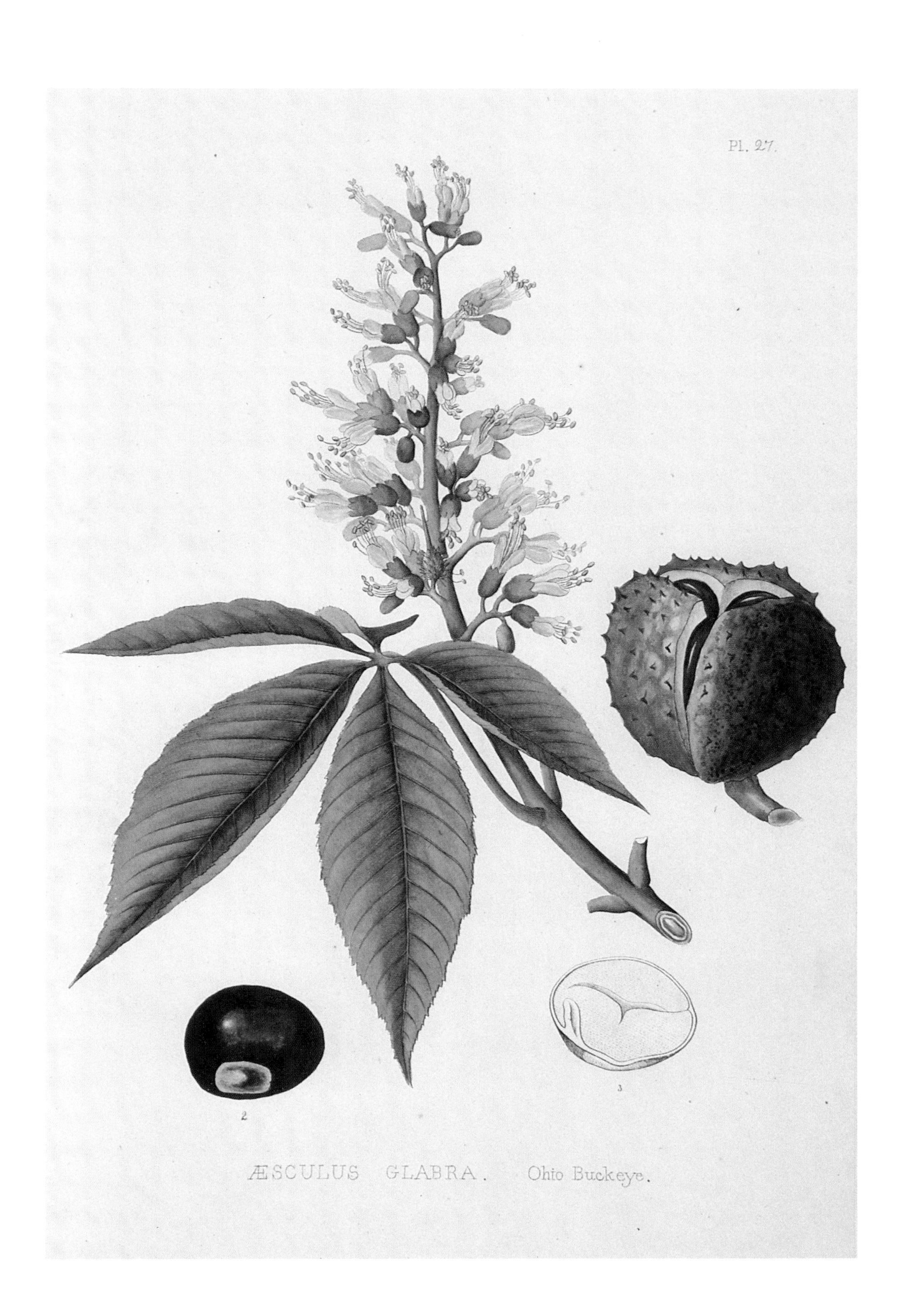
Pl. 27.
2
3
ÆSCULUS GLABRA. Ohio Buckeye.

Plate 34 *Mountain Maple*

Acer Spicatum. Mountain Maple
Unsigned lithograph; drawn by Isaac Sprague;
engraved on stone, printed, and colored by Joseph Prestele; 34 × 27 cm.
Plate 25 for the projected *Forest Trees of North America*, by Asa Gray.

Joseph Prestele wrote Asa Gray on February 11, 1851, that he would make "the shades in the fruits behind the leaf more light."

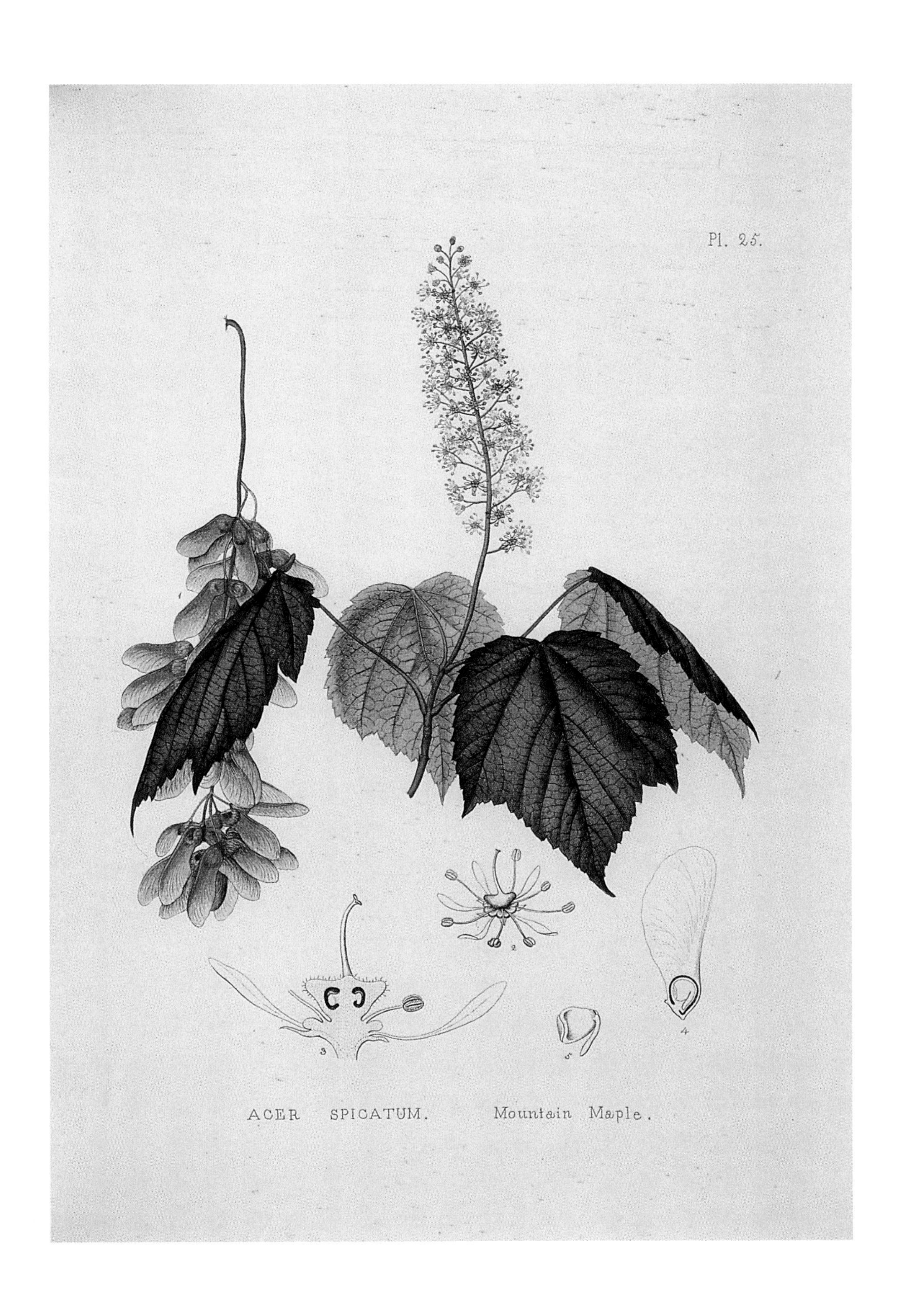
Pl. 25.
2
3
5
4
ACER SPICATUM. Mountain Maple.

Plate 35 *Red Maple*

Acer Rubrum. *Red Maple*
Unsigned lithograph; drawn by Isaac Sprague;
engraved on stone, printed, and colored by Joseph Prestele; 34 × 27 cm.
Plate 20 for the projected *Forest Trees of North America*, by Asa Gray.

Gray criticised the colored proof of this plate that Prestele sent him, and in his reply on February 11, 1851, the artist wrote that he could "improve and colour the flowers more light, which will render them to your satisfaction." He added that he had increased his charge for coloring this plate to 10 cents each.

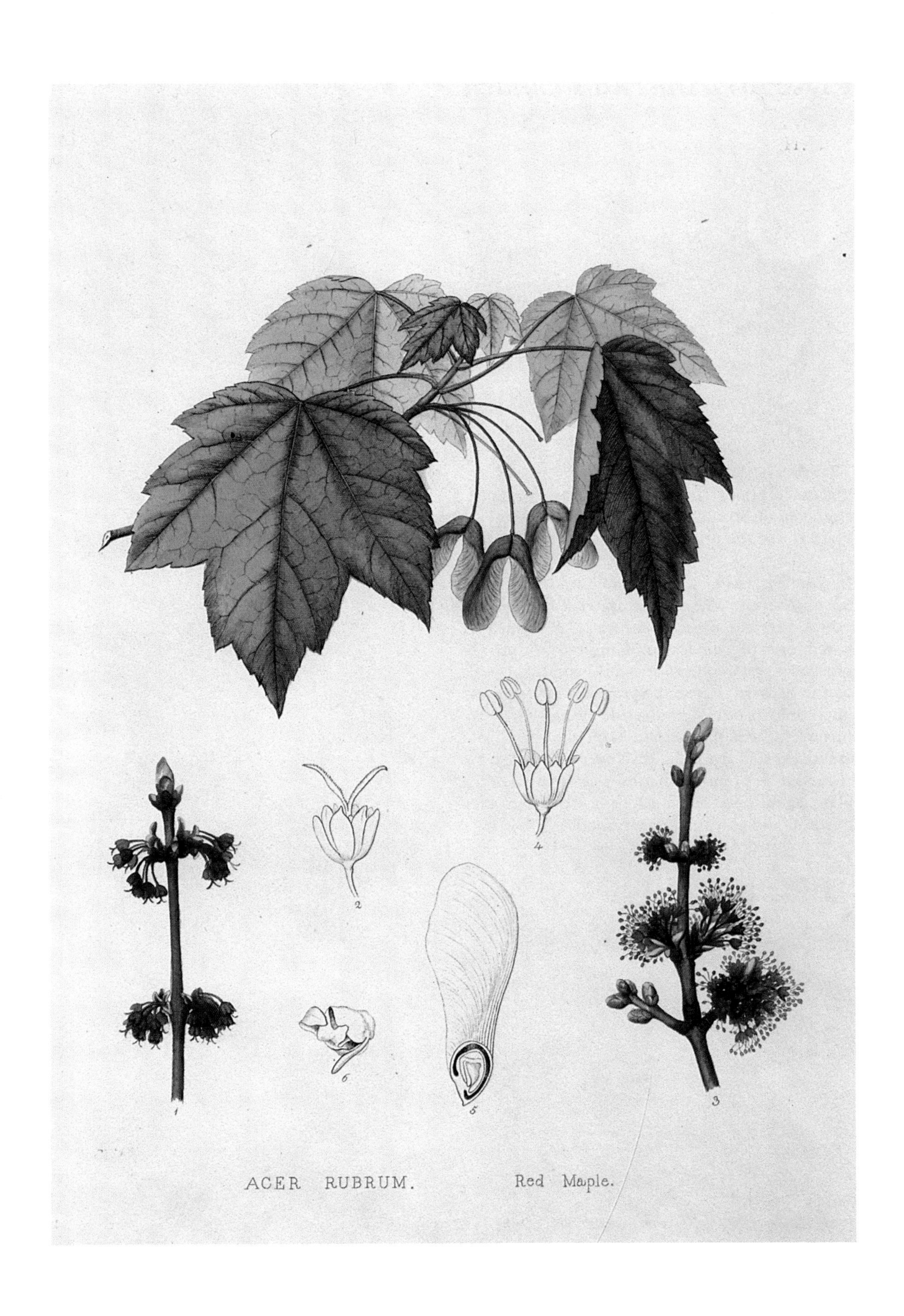

ACER RUBRUM. Red Maple.

Plate 37 *Tulip Tree*

Liriodendron Tulipifera. Tulip Tree
"Sprague ad nat. del."
Lithograph; engraved and colored by Joseph Prestele;
printed by Tappan & Bradford, Boston, Mass.; 34 × 27 cm.
Plate 8 for the projected *Forest Trees of North America*, by Asa Gray.

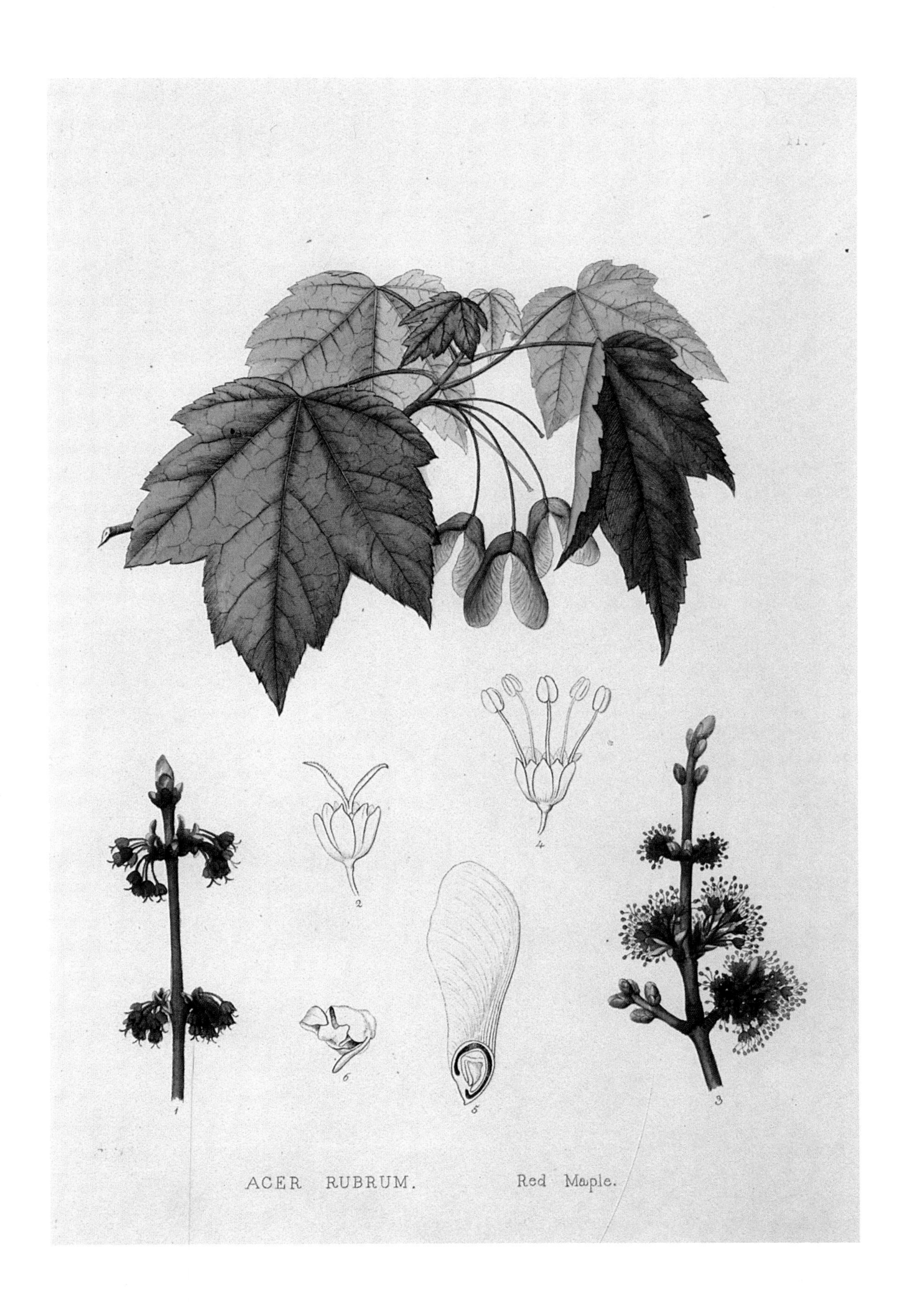
1
2
3
4
5
6
ACER RUBRUM. Red Maple.

Plate 36 *American Linden*

Tilia Americana. American Linden
Unsigned lithograph; drawn by Isaac Sprague;
engraved on stone, printed, and colored by Joseph Prestele; 34 × 27 cm.
Plate 10 for the projected *Forest Trees of North America*, by Asa Gray.

In January 1851, while Gray was in Europe, Sprague sent to Prestele the drawing for his plate which Prestele altered because he thought that when one of the leaves, hanging behind the blossoms, was colored it would spoil the looks of the flowers. He sent Sprague a colored and uncolored proof of the result. When Gray returned he was dissatisfied with the proof and asked Prestele to redo it. The repentant artist replied in September that he would engrave the plate again and "take all pains to do it very graceful and to your full satisfaction."

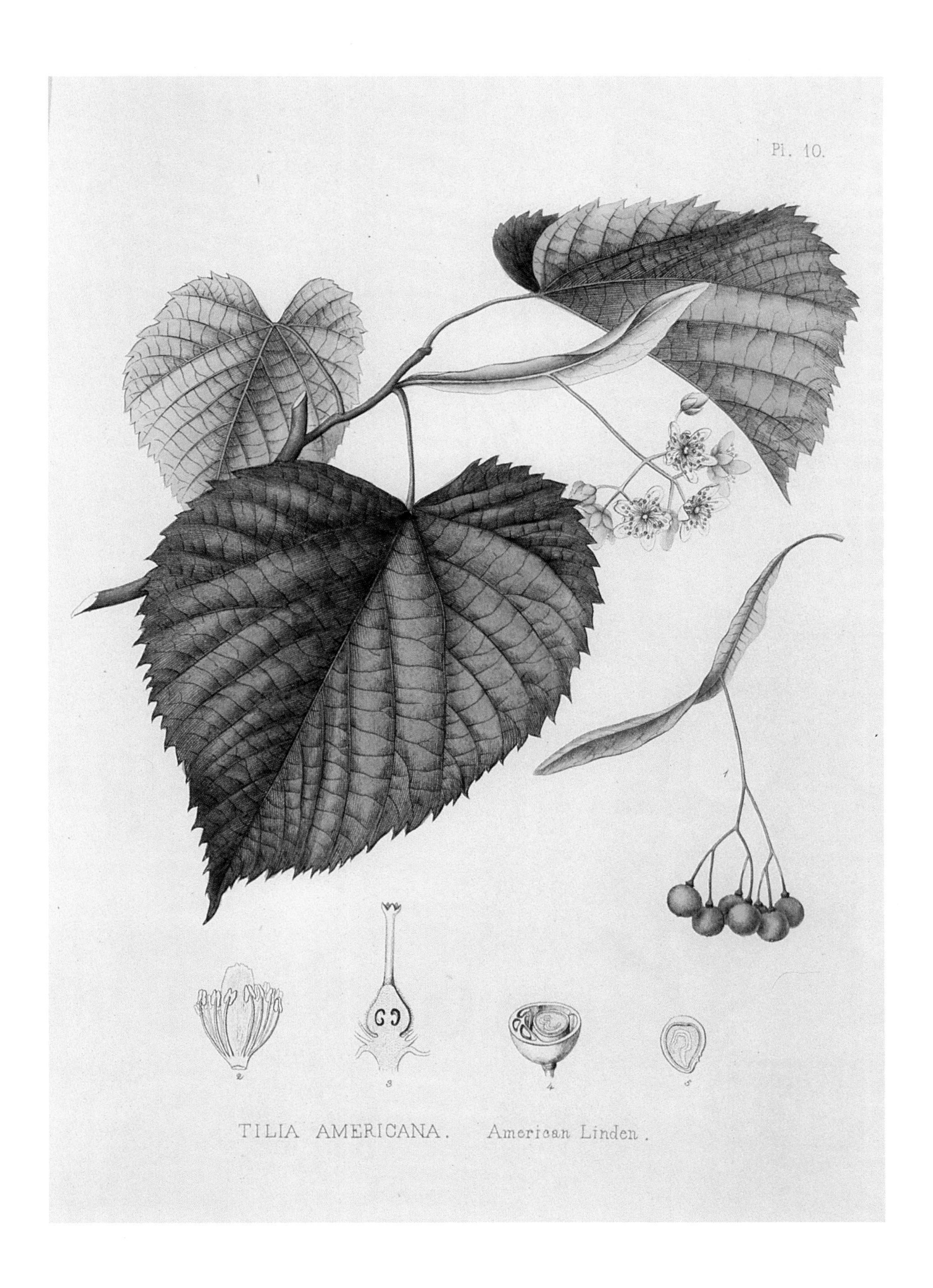

TILIA AMERICANA. American Linden.

Plate 37 *Tulip Tree*

Liriodendron Tulipifera. Tulip Tree
"Sprague ad nat. del."
Lithograph; engraved and colored by Joseph Prestele;
printed by Tappan & Bradford, Boston, Mass.; 34 × 27 cm.
Plate 8 for the projected *Forest Trees of North America*, by Asa Gray.

Pl. 8
2.
3.
4.
1.
Sprague ad nat del.
LIRIODENDRON TULIPIFERA
Tulip Tree.

Plate 38 *Dogwood*

Cornus Alternifolia. Dogwood
(*Cornus alternifolia.* Pagoda Dogwood, Alternate leaf Dogwood, Blue Dogwood, etc.)
"Sprague del."
Lithograph; engraved, printed, and colored by Joseph Prestele; 34 × 27 cm.
Plate 63 for the projected *Forest Trees of North America*, by Asa Gray.

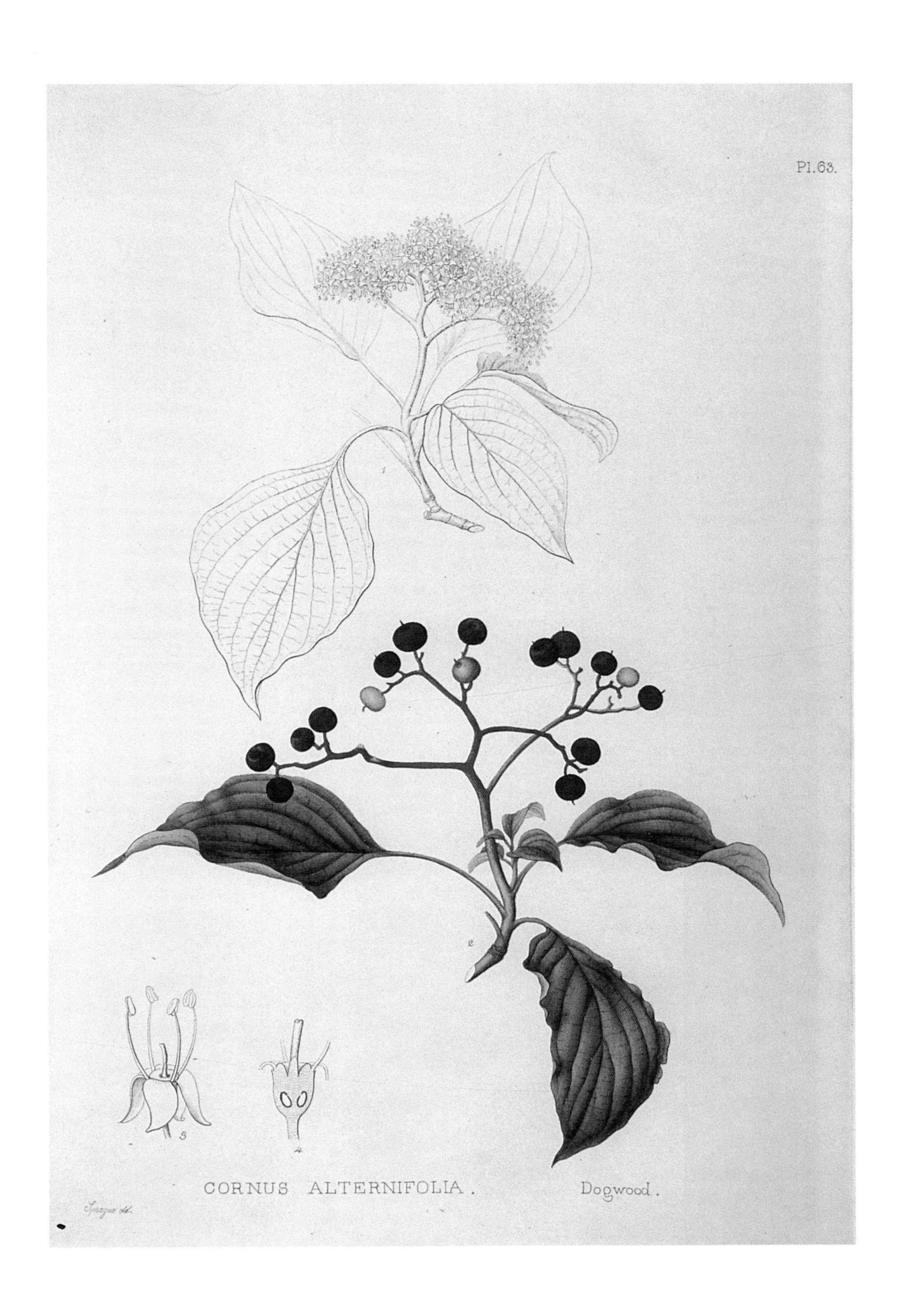
Pl.63.
CORNUS ALTERNIFOLIA. Dogwood.

Plate 39 *Bouquet of Chrysanthemums*

Bouquet of Chrysanthemums
"1. Bouquet parfaite. 2. La Vogue. 3. Selfabare. 4. Valeda. 5. Bernettianum.
6. Helen. 7. La Rousse. 8. Grand Sultan."
Lithograph; attributed to Joseph Prestele; engraved on stone, colored; 25.5 × 16 cm.
From the *Horticulturist*, n.s., vol. 5, no. 3 (March 1855), colored edition.
Courtesy, Elisabeth Woodburn, Hopewell, New Jersey.

Plates 39–41

In 1853 James Vick, a Rochester, New York, publisher soon to become one of America's best-known seedsmen, bought the *Horticulturist, A Journal of Rural Art and Rural Taste*, following the tragic death of its founder and editor, Andrew Jackson Downing. Vick moved the influential journal to Rochester and engaged Patrick Barry, a partner in the Mount Hope Nursery of Rochester and a cultivated man of many abilities, to serve as its editor during the two years of Vick's ownership. Among Vick's innovations was to add an edition of the magazine with colored illustrations. Many of these plates were engraved and carefully handcolored by Joseph Prestele and were of the highest quality. These provided an added element of beauty and interest to the journal, whose reputation was ably maintained by Barry. Many of the Prestele illustrations were of fruit, particularly pears which were then of great interest to horticulturists. In addition, the artist created plates of flowers recently introduced to American gardens, including some brought from the Orient. Unfortunately, the artist had some unhappy experience with Vick which left him bitter, at least as bitter as Prestele, with his philosophical acceptance of life's hard knocks, could be.

1 Bouquet parfaite. 2 La Vogue. 3 Solfatare. 4 Valeda.
5 Bernettianum. 6 Helen. 7 La Rousse. 8 Grand Sultan.

Plate 40 *Double White Flowering Almond Double Crimson Flowering Peach*

Double White Flowering Almond. Double Crimson Flowering Peach
(*Prunus* [*glandulosa 'albiplena'?*]. *Prunus persica 'Rubroplena.'*)
Unsigned lithograph; attributed to Joseph Prestele;
engraved on stone, colored; 23.5 × 15.5 cm.
From the colored edition of the *Horticulturist*,
n.s. vol. 5, no. 5 (May 1855).

Like many of Joseph Prestele's most popular and oft-printed plates, this design went through a number of reprints and revisions. The order of that progression isn't clear. What may be the first is a nurserymen's plate, unsigned. Then James Vick may have ordered copies of it to use as illustrations in both the uncolored and colored editions of his *Horticulturist*, the plate reproduced here. It also seems likely that D. M. Dewey bought reprints of the plate to issue as a nurserymen's plate with his—Dewey's—imprint, as one is included in *The Colored Fruit Book for the Use of Nurserymen* (Rochester, New York, 1859). A later version of this same subject, showing the sprays intertwined, also attributed to Joseph Prestele, appears to date after the artist moved to Amana, Iowa, in 1858, as the only copies seen were in collections there (see Plate 78).

Prestele's infinite care in finishing his plates is shown in the one reproduced here. The tiny details in the blossoms were added to each plate in pencil.

Double white flowering
Almond.
Shrub Var.
Double crimson flowering
Peach.
Tree Var.

Plate 41 *Dielytra Spectabilis*

Dicentra or *Dielytra Spectabilis*
(*Dicentra spectabilis.* Common Bleedingheart)
Lithograph; attributed to Joseph Prestele;
engraved on stone, colored; 23.5 × 16 cm.
From the colored edition of the *Horticulturist*,
vol. 4, no. 7(July 1854), following page 296.
Courtesy, Ellwanger and Barry Collection,
Department of Rare Books and Special Collections, Rush Rhees Library, University of Rochester.

Joseph Prestele produced a nurserymen's plate of this subject as well; a variant plate was also produced by his son William Henry.

Robert Fortune, the plant collector, brought the *Dicentra spectabilis* from China to England in 1846. It first flowered in America in 1851 where, because of its beauty and ease of propagation, it soon became a familiar and much beloved garden plant. It acquired its popular name in England to become another of several flowers called "Bleeding Heart."

a
b
c
DICENTRA or DIELYTRA SPECTABILIS.

Plate 42 *Blue Pearmain Apple*

The Blue Pearmain Apple
"By J. & G. Prestele. Amana, Iowa"
Watercolor; caption and signature handwritten; 32 × 24 cm.
Courtesy, Ellwanger and Barry Collection,
Department of Rare Books and Special Collections,
Rush Rhees Library, University of Rochester.

Plates 42–83

In the late 1840s Joseph Prestele, aided by his son Gottlieb, began making and selling to nurserymen and their traveling agents beautiful colored prints illustrating the fruits, flowers, and ornamental shrubs and trees which the nurseries had for sale. Prestele based the designs of his plates upon horticultural and botanical illustrations which he simplified and adapted to the commercial needs of nurserymen. The plates were all lithographed, some being printed from drawings in "chalk" on stone, others from engravings cut in stone in fine, delicate lines. All were patiently handcolored. The result was a new form of commercial art, uniquely American.

The Presteles did not advertise their product, but the plates gradually became known as salesmen found them effective in increasing sales. In time the plate business provided the Presteles a modest but steady source of income.

When Prestele introduced his plates, the time was right, for nurseries were growing in number and size because the nation was growing, and new gardens and new orchards were being planted. Steamboats, canals, and railroads made it easier to ship plant material to distant points. Nursery salesmen visited rural areas in many parts of America and Canada soliciting sales, and took their plate books with them; the use of mail-order catalogues also increased.

Prestele's modest success was noted with interest by enterprising publishers and others, and by the late 1850s competitors had entered the field.

After the Civil War, Joseph Prestele's youngest son, William Henry, who had left home when he was young, took up making nurserymen's plates and continued making them into the 1880s, long after his father's death and after his brother Gottlieb had retired from that work. William Henry's plates are very unlike those his father and brother created, less skilled but more innovative in design, and bolder in pattern and color.

Toward the end of the nineteenth century, nurserymen's plates gradually became obsolete as inexpensive color printing was developed, and lavishly and richly colored catalogues could be produced cheaply and distributed in huge quantities.

The Blue Permain Apple.
By J. & G. Paetzle. Amana. Iowa.

Plate 43 *Swaar Apple*

The Swaar Apple
Unsigned lithograph; attributed to Gottlieb Prestele; chalk on stone; colored; 34 × 24 cm.
Courtesy, Mrs. Joseph Mattes, Amana, Iowa.

The Swaar apple originated among the Dutch settlers on the Hudson River, near Esopus, New York, sometime before 1800. Its name means "heavy" or "large" in Low Dutch—fruit twelve inches in circumference have been reported. It is yellowish green when ripe, of "uncommon flavour and richness," as William Coxe of New Jersey wrote of it in his *View of the Cultivation of Fruit Trees* (1817). Andrew Jackson Downing described it as a "truly noble American fruit," its flesh fine grained and tender, "with an exceedingly rich aromatic flavour, and a spicy smell." By 1905 only "a few trees are still found in old orchards," S. A. Beach wrote in his *Apples of New York*. "It has nowhere been cultivated extensively and is now seldom planted."

THE SWAAR APPLE.

Plate 44 *Summer Rose Apple*

The Summer Rose Apple
Unsigned lithograph; attributed to Joseph Prestele; colored; caption handwritten; 33.5 × 26.5 cm. Courtesy, Mrs. Emma Setzer, South Amana, Iowa.

This old variety originated in New Jersey. The fruit is of singular beauty, as the nurseryman William Coxe wrote in 1817, of delicate texture and clear waxy skin, qualities that resulted in one of its many names being "Glass." It was excellent both for eating and stewing although rather small for popular use. Commercial production of the variety gradually ceased about 1900.

The Summer Rose Apple.

Plate 45 *Siberian Crab Apple*

I Small Red II Black Siberian Crab Apple
Unsigned lithograph; attributed to Gottlieb Prestele;
drawn in chalk on stone, colored; 34 × 26.5 cm.
Courtesy, History of Science Collections,
Cornell University Libraries, Ithaca, New York.

I SMALL RED II BLACK

SIBERIAN CRAB APPLE.

Plate 46 *White French Guigne Cherry*

The White French Guigne Cherry
Unsigned lithograph; attributed to Joseph Prestele;
engraved on stone, colored; 32 × 24 cm.
Courtesy, Ellwanger and Barry Collection,
Department of Rare Books and Special Collections,
Rush Rhees Library, University of Rochester.

THE WHITE FRENCH GUIGNE CHERRY.

Plate 47 *Tradescant's Black Heart Cherry*

The Tradescant's Black Heart Cherry
Unsigned lithograph; attributed to Gottlieb Prestele;
drawn in chalk on stone, colored; 34 × 26.5 cm.
Courtesy, Ellwanger and Barry Collection,
Department of Rare Books and Special Collections,
Rush Rhees Library, University of Rochester.

THE TRADESCANT'S BLACK HEART CHERRY.

Plate 48 *Crawford's Late Peach*

Crawford's Late Peach
Unsigned lithograph; attributed to Gottlieb Prestele; drawn in chalk on stone, colored; 34 × 26.5 cm.
Courtesy, History of Science Collections, Cornell University Libraries, Ithaca, New York.

Crawford's late peach was a prized American variety when Andrew Jackson Downing rhapsodized in his *Fruits and Fruit Trees of America* (1845) that it was one of the most magnificent American peaches, and deserving of universal cultivation. It was a freestone, with yellow flesh, juicy, melting, with a very rich and excellent vinous flavor, ripening during the last days of September. It originated with William Crawford, Middletown, New Jersey, who also created a famous early ripening peach.

CRAWFORD'S LATE PEACH.

Plate 49 *Madeline*

Madeline
"G. Prestele"
Lithograph; drawn in chalk on stone, colored; 34 × 26.5 cm.
From an office file copy, Mount Hope Nursery
(Ellwanger & Barry), Rochester, New York.
Courtesy, Ellwanger and Barry Collection,
Department of Rare Books and Special Collections,
Rush Rhees Library, University of Rochester.

Gottlieb Prestele had problems with the spelling of this caption. A copy of this plate in the Cornell University collection is captioned "Madelaine."

MADELINE.

Plate 50 *Beurre d' Aremberg Pear*

The Beurre d'Aremberg Pear
"Lith. & col[d]. by Amana Society. Amana, Iowa County, Iowa"
Lithograph; drawn in chalk on stone, colored; 34 × 26.5 cm.
From an office file copy, Mount Hope Nursery
(Ellwanger & Barry), Rochester, New York.
Courtesy, Ellwanger and Barry Collection,
Department of Rare Books and Special Collections,
Rush Rhees Library, University of Rochester.

THE BEURRE D'AREMBERG PEAR.

Lith. & colᵈ. by Amana Society. Amana, Iowa County, Iowa.

Plate 51 *Virgalieu Pear*

The Virgalieu or *White Doyenne Pear*
"Lith. & col^{d}. by Amana Society. Amana, Iowa County, Iowa"
Lithograph; attributed to Gottlieb Prestele;
drawn in chalk on stone, colored; 34 × 24.5 cm.
From an office copy, *Album of Fruit*, vol. 2,
Mount Hope Nursery (Ellwanger & Barry),
Rochester, New York.
Courtesy, Ellwanger and Barry Collection,
Department of Rare Books and Special Collections,
Rush Rhees Library, University of Rochester.

THE VIRGALIEU or WHITE DOYENNE PEAR.

Plate 52 *Rapalje Seedling Pear*

The Rapalje Seedling Pear
"By J. & G. Prestele. Amana Society, Iowa"
Watercolor(?); 34 × 27.8 cm.
From an unbound office copy, Mount Hope Nursery (Ellwanger & Barry), Rochester, New York.
Courtesy, Ellwanger and Barry Collection, Department of Rare Books and Special Collections, Rush Rhees Library, University of Rochester.

The Rapaljie Seedling Pear
By J & G. Prestele
Amana Society. Iowa.

Plate 53 *Pratt Pear*

The Pratt Pear
"By J. & G. Prestele. Amana, Iowa Co. Iowa"
Watercolor; 34 × 27.8 cm.
From an office file copy, Mount Hope Nursery (Ellwanger & Barry), Rochester, New York.
Courtesy, Ellwanger and Barry Collection, Department of Rare Books and Special Collections, Rush Rhees Library, University of Rochester.

The Pratt Pear
By J. & G. Prestele
Amana Iowa Co. Iowa

Plate 54 *Lawrence's Favorite Plum*

The Lawrence's Favorite Plum
Unsigned lithograph; attributed to Gottlieb Prestele; drawn in chalk on stone, colored; 28 × 21.5 cm. Courtesy, Ellwanger and Barry Collection, Department of Rare Books and Special Collections, Rush Rhees Library, University of Rochester.

F. L. Lawrence, Hudson, New York, planted a seed of the Green Gage plum sometime during the early part of the nineteenth century, and produced this plum which resembles its parent but is several times larger. The skin is, as described by Andrew Jackson Downing, a dull yellowish green, clouded with streaks of a darker shade beneath, and covered with a light bluish green bloom. The flesh is remarkably juicy and melting, with a very rich, sprightly, vinous flavor, and one of the most delicious of plums. It ripens at the middle of August. Downing felt that its many virtues would soon give it a place in every garden.

THE LAWRENCE'S FAVORITE PLUM.

Plate 55 *Nebraska Seedling Plum* *Thompson's Golden Gem Plum*

1. The Nebraska Seedling Plum. 2. Thompson's Golden Gem Plum
"Introduced by R. O. Thompson, Nursery Hill, Nebraska"
"Lith. & col[d]. by Amana Society. Amana, Iowa County, Iowa"
Lithograph; drawn in chalk on stone; colored; 27.8 × 21.8 cm.
From an office file copy, Mount Hope Nursery
(Ellwanger & Barry), Rochester, New York.
Courtesy, Ellwanger and Barry Collection,
Department of Rare Books and Special Collections,
Rush Rhees Library, University of Rochester.

1.
2.
2.
1. THE NEBRASKA SEEDLING PLUM. 2. THOMPSON'S GOLDEN GEM PLUM.
Introduced by R. O. Thompson, Nursery Hill, Nebraska.
Lith. & cold. by Amana Society.
Amana, Iowa County, Iowa.

Plate 56 *Mc Laughlin Plum*

The Mc Laughlin Plum
"J. Prestele. Ebenezer, n. Buffalo"
Lithograph; drawn on chalk on stone; colored; 28 × 22.5 cm.
From a bound assortment of plates issued by the
Mount Hope Nursery (Ellwanger & Barry) Rochester, New York,
to one of their agents in the spring of 1871.
Author's collection.

"Very sweet, and luscious" was how Andrew Jackson Downing described this plum. The fruit was large, nearly round, and the tender skin yellow, dotted and marbled with red on the sunny side, and covered with a thin bloom, as shown in this plate.

The plum originated with James McLaughlin, Bangor, Maine.

THE MC LAUGHLIN PLUM.

Plate 57 *Chasselas Blanc Grape*

Chasselas Blanc Grape
Unsigned lithograph; attributed to Gottlieb Prestele;
drawn in chalk on stone, colored; 34 × 26.5 cm.
Courtesy, History of Science Collections,
Cornell University Libraries, Ithaca, New York.

CHASSELAS BLANC GRAPE.

Plate 58 *Concord Grape*

The Concord Grape
Unsigned lithograph; attributed to Gottlieb Prestele; drawn in chalk on stone, colored; 34 × 26.5 cm.
Courtesy, History of Science Collections, Cornell University Libraries, Ithaca, New York.

The Concord grape has been the most popular and most widely grown grape in America—particularly in family gardens—since it was developed by Ephraim W. Bull of Concord, Massachusetts, after years of patient experimentation. He first exhibited it in September 1853 at the annual exhibition of the Massachusetts Horticultural Society in Boston. It is sad to note that Bull, this public benefactor, lost everything in his efforts to promote his grape and in his last years was supported by the charity of friends. He died, aged eighty-seven, from a fall from a ladder while working in his garden.

THE CONCORD GRAPE.

Plate 59 *White Grape*

White Grape
(Currant)
Unsigned lithograph; attributed to Gottlieb Prestele; drawn in chalk on stone, colored; 34 × 24.5 cm.
Courtesy, Ellwanger and Barry Collection, Department of Rare Books and Special Collections, Rush Rhees Library, University of Rochester.

WHITE GRAPE.

Plate 60 *Cherry Currant*

The Cherry Currant
Unsigned lithograph; attributed to Gottlieb Prestele; drawn in chalk on stone, colored; 34 × 24.5 cm.
Courtesy, Ellwanger and Barry Collection, Department of Rare Books and Special Collections, Rush Rhees Library, University of Rochester.

THE CHERRY CURRANT.

Plate 61 *Yellow Raspberry*

The Yellow Raspberry
Unsigned lithograph; attributed to Gottlieb Prestele;
drawn in chalk on stone, colored; 30 × 23.7 cm.
Courtesy, Mrs. Adolph Schmieder, Middle Amana, Iowa.

THE YELLOW RASPBERRY.

Plate 62 *Whitesmith Gooseberry (Woodward's)*

The Whitesmith Gooseberry. (Woodward's)
"Lith. & col^{d}. by Amana Society, Homestead, Iowa"
Unsigned lithograph; attributed to Gottlieb Prestele;
drawn in chalk on stone, colored; 30 × 21.5 cm.
Courtesy, Mrs. Adolph Schmieder, Middle Amana, Iowa.

THE WHITESMITH GOOSEBERRY.
(WOODWARD'S.)

Lith & col'd by Amana Society, Homestead Iowa.

Plate 63 *Trollopes Victoria*

Trollopes Victoria
(Strawberry)
Unsigned lithograph; attributed to Joseph Prestele;
engraved on stone, freehand details; 28 × 22.5 cm.
From a bound assortment of plates issued by the
Mount Hope Nursery (Ellwanger & Barry), Rochester, New York,
to one of their agents in the spring of 1871.
Author's collection.

A number of plates of strawberries, closely related in design to this one, are also attributed to Joseph Prestele before he moved to Iowa in 1858. At Amana, Prestele produced a plate captioned "Boston Pine Strawberry" whose design is a variation of those he made earlier in New York State.

Trollope's Favorite, also known as Victoria, Golden Queen, Trembly's Union, and, simply, Union, was an English variety. The fruit was a light crimson color with tender, juicy, sweet, scarlet flesh, and a "somewhat peculiar aromatic flavor." American growers reported it hardy but not very productive. It too, like other early varieties, had a limited span of popularity before being replaced by newer varieties. Few fruits are as evanescent as cultivated varieties of strawberries, or as dependent upon the propagation and protection of man.

TROLLOPES VICTORIA

Plate 64 *Bicton Pine*

Bicton Pine
(Strawberry)
Unsigned lithograph; attributed to Joseph Prestele;
engraved on stone; freehand details; colored; 32 × 24.5 cm.
From an office copy, *Album of Fruit*, vol. 2, Mount Hope Nursery
(Ellwanger & Barry), Rochester, New York.
Courtesy, Ellwanger and Barry Collection,
Department of Rare Books and Special Collections,
Rush Rhees Library, University of Rochester.

The Bicton Pine—sometimes called the Virgin Queen—was an English variety introduced to American gardens during the late 1850s. It was, as can be seen in this colored plate, unusual. It didn't *look* like a strawberry, at least a ripe one, for the fruit was white with only a faint blush on the side exposed to the sun. Writers politely stated that it was quite a novelty. They spoke of its flavor as being "delicate, mild and pleasant, but not rich," and they conceded that it was more productive than might be expected but somewhat tender. They suggested that it would probably only find a place in amateurs' gardens in limited quantities. It would seem that these American horticulturists didn't take the Bicton Pine very seriously.

BICTON PINE.

Plate 65 *Double Flowering Sweet Violets. Snow Drops*

1 Double Flowering Sweet Violets. *2 Snow Drops*
Unsigned lithograph; attributed to Joseph Prestele;
engraved on stone, colored; 26.5 × 21.5 cm.
Courtesy, Mrs. Emma Setzer, South Amana, Iowa.

1 DOUBLE FLOWERING SWEET VIOLETS. 2. SNOW DROPS.

Plate 66 *Tulips*

Tulips
Unsigned lithograph; attributed to Gottlieb Prestele; drawn in chalk on stone, colored; 28.3 × 22.2 cm.
Author's collection.

This plate clearly shows the style of drawing used by Gottlieb during his later years; it is freer and made with more shading than his earlier work which followed closely his father's meticulous style using very fine lines. Prestele reworked an earlier version, *Tulip* (shown in black-and-white), by adding a single "bizarre" [striped] tulip to create the plate reproduced here.

Tulip
Unsigned lithograph; attributed to Gottlieb Prestele; drawn in chalk on stone; 32 × 23.5 cm.
Courtesy, Amana Society, Amana, Iowa

TULIPS.

Plate 67 *Chinese Poeonias*

Chinese Poeonias
"1. P. Pottsii. 2. P. Carnea. 3. P. Compte de Paris."
Unsigned lithograph; attributed to Joseph Prestele; drawn in chalk on stone; 30.5 × 23.5 cm.
Author's collection.

Courtesy, Mrs. Joseph Mattes, Amana, Iowa.

CHINESE PŒONIAS.

1 P. Pottsii. 2 P. Carnea. 3 P. Compte de Paris.

Plate 68 *Lilium Lancifolium Rubrum*

Lilium Lancifolium Rubrum
(*Lilium speciosum 'Rubrum.'* Pink Speciosum Lily)
Unsigned lithograph; attributed to Joseph Prestele;
drawn in chalk on stone, colored; 28 × 22.5 cm.
Courtesy, Ellwanger and Barry Collection,
Department of Rare Books and Special Collections,
Rush Rhees Library, University of Rochester.

A variant of this design, captioned "Lilium Lancifolium Speciosa," including only a single blossom with a bud and leaves, was made by Joseph Prestele as an illustration in the *Horticulturist*, March 1854.

This lily was introduced to European gardens by the Bavarian naturalist and traveler, Philipp Franz von Siebold (1796–1866) who collected the plant while on a Dutch mission to Japan, 1823–30. After von Siebold returned to Europe and began compiling his *Flora Japonica* Joseph Prestele was one of the artists engaged to make many of the plates, examples of which are reproduced in Plates 4, 5. Possibly the artist first learned of this lily at that time.

When the lily began to be grown in Europe, it was acclaimed in *Paxton's Magazine of Botany* (London, January 1838) as a "most magnificent species." The high cost of the bulbs limited its use in American gardens for some years, but by the mid-1850s the prices had become "reasonable and within the reach of persons of moderate means," causing a writer to predict that "the Crimson Lance-leaved Japan Lily" would become very popular, as it quickly did.

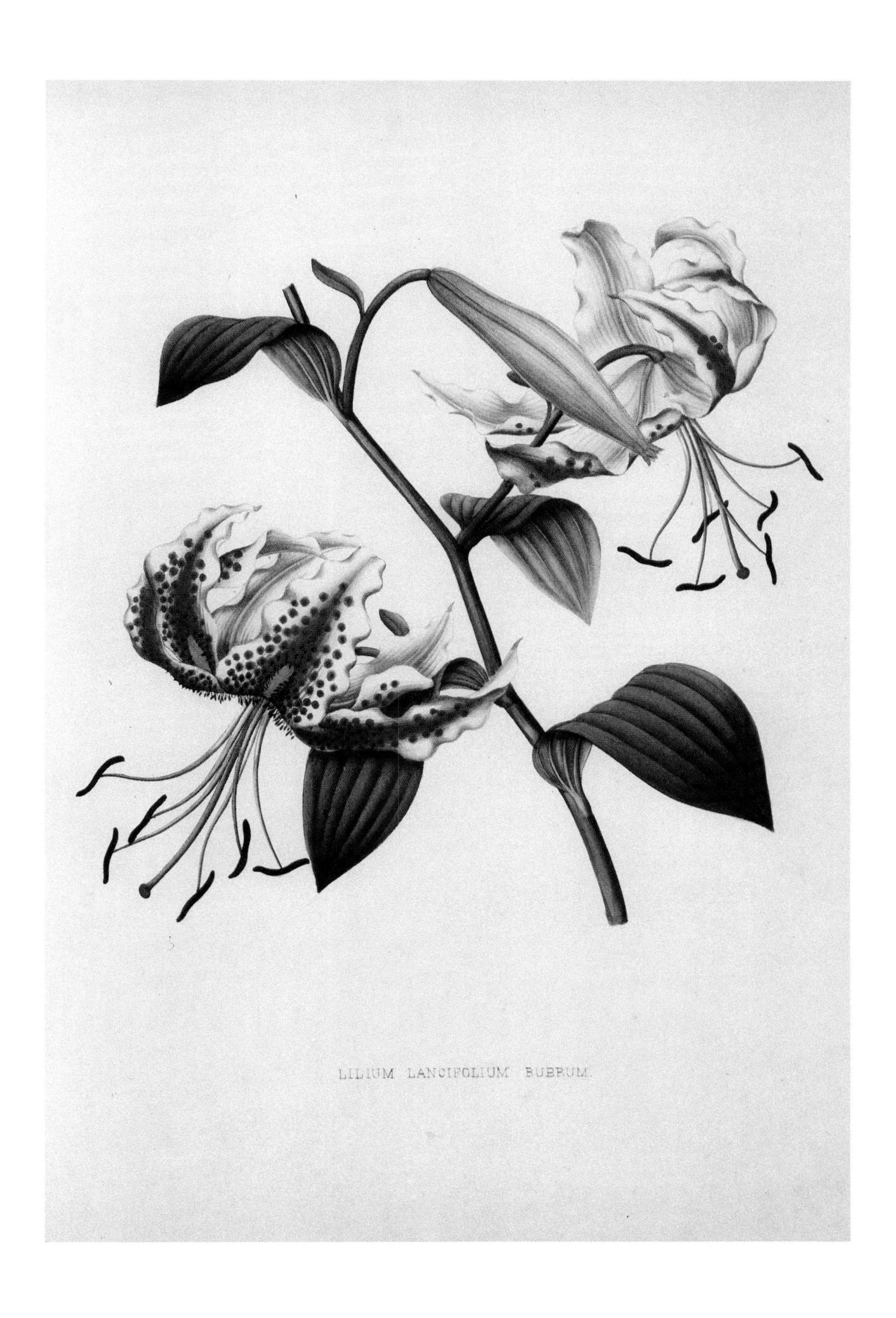

LILIUM LANCIFOLIUM RUBRUM.

Plate 69 *Chrysanthemums*

Chrysanthemums
"1 La Gitano 2 Criterion 3 Asmodea 4 Daphnis 5 Perfecta."
Unsigned lithograph; attributed to Joseph Prestele;
engraved on stone, colored; 32 × 26.7 cm.
Courtesy, Ellwanger and Barry Collection,
Department of Rare Books and Special Collections,
Rush Rhees Library, University of Rochester.

This print, issued as a nurserymen's plate, is a simplified version of an illustration published in the *Horticulturist* (n.s., vol. 3, no. 3, March 1853) showing the same varieties and with the caption beautifully engraved in script.

CHRYSANTHEMUMS,

1 La Gitano 2 Criterion 3 Asmodea 4 Daphnis 5 Perfecta.

Plate 70 *Moss Rose*

Moss Rose
Unsigned lithograph; attributed to Joseph Prestele;
engraved on stone, colored; 28 × 21 cm.
From a nurserymen's plate book assembled by the
Mount Hope Nursery (Ellwanger & Barry), Rochester, New York,
for one of its agents in the spring of 1871.
Author's collection.

In the language of flowers, the Moss rose symbolized "Pleasure without alloy."

"There is no rose that has been, and is still so highly esteemed as the Moss," commented Robert Buist, Philadelphia nurseryman, in 1844. "It is figured and emblazoned in every quarter of the globe . . . and among them all it is questionable if there is one so very beautiful in bud as the common Moss Rose, generally known under the name of Red Moss, in contradistinction, I suppose, to white, for it is not red; it is purely rose-colour, and in bud is truly lovely, but when full blown it has no peculiar attraction." (Buist, *The Rose Manual*, 1844).

Writers spoke of the fragrance of the bloom, its effectiveness in mass plantings, its heavy foliage, and its uncomplaining hardiness. But it was the moss-covered buds that fascinated, and which were so different from any others.

The Moss rose probably developed in Holland as a sport of the ancient Provence rose and was taken to England during the sixteenth century. American gardeners like George Washington grew it during the 1700s. For many years only a few varieties existed but by the 1840s more than a hundred new forms had been developed including white ones. A few, like Glory of the Mosses, achieved some popularity but the old-fashioned Common Moss continued to be appreciated as one of the most beautiful of the whole family. Andrew Jackson Downing said that if he could have only three roses for his personal pleasure, one would be "Old Moss Rose." (*Horticulturist*, vol. 3, no. 2 [August 1848], p. 62.)

MOSS ROSE.
Perpetual var.

Plate 71 *Gen. Jacqueminot Rose*

Gen. Jacqueminot Rose
Unsigned lithograph; attributed to Joseph Prestele; engraved on stone, colored; 22.5 × 16.4 (trimmed). From an office file copy, Mount Hope Nursery (Ellwanger & Barry), Rochester, New York. Courtesy, Ellwanger and Barry Collection, Department of Rare Books and Special Collections, Rush Rhees Library, University of Rochester.

In 1866 Francis Parkman, who was both a horticulturist and a historian, wrote with unusual enthusiasm about the General Jacqueminot as being "one of the most splendid of roses." The blossoms of this Hybrid Perpetual were fragant, "rich and velvety" crimson, and, under good cultivation, grew to immense size. They glowed, so Parkman thought, "like a firebrand among the paler hues around it." The plant was hardy, a strong grower, and was recommended both for garden planting and for forcing. Because of its many admirable qualities, plant breeders used it as the ancester of a numerous progeny.

General Jacqueminot was developed by Roussel at Montpelier, France, in 1853. Ellwanger & Barry in their 1881 catalogue speculated that it was a seedling of the Hybrid China rose, Gloire des Rosomanes. Indeed, a purplish crimson Hybrid Chinese rose was also named General Jacqueminot.

The crimson variety shown here was propagated by nurseries into the 1930s.

GEN: JACQUEMINOT ROSE.

Plate 72 *Augusta Rose*

The Augusta Rose
Unsigned lithograph; attributed to Joseph Prestele; engraved on stone, colored; 34 × 26.5 cm.
From an office file copy, Mount Hope Nursery (Ellwanger & Barry), Rochester, New York.
Courtesy, Ellwanger and Barry Collection, Department of Rare Books and Special Collections, Rush Rhees Library, University of Rochester.

Joseph Prestele apparently made an engraved lithograph of the Augusta rose sometime before 1859 because what seems to be a version of it was published by D. M. Dewey in his 1859 *Colored Fruit Book*. Thereafter Joseph, perhaps his son Gottlieb, and certainly William Henry Prestele, created nurserymen's plates illustrating it.

Few plants have been introduced with as much excitement or suspense or have created as much controversy as this rose. The story is this, as recounted by A. Fahnestock, Lancaster, Ohio, July 1849—a friend of the principals—to the editor of the *Horticulturist* (vol. 4, no. 3, September 1849, pp. 147–48).

On a balmy January day in 1844, the Honorable James Matthews of Coshocton, Ohio, with his wife, and the Honorable A. P. Stone, and Mrs. Stone of Columbus, Ohio, visited Mount Vernon. While there, Matthews, a rose fancier, gathered seed from some of the roses growing in what had been George Washington's garden. These he planted at his Ohio home and one of them produced a vigorous plant which in two seasons had sent up shoots sixteen to eighteen feet long. The blossoms—as Matthews reported to Fahnestock—"were very large and very double, and in colour a light, pure yellow . . . much larger and much more double than those on the Chromatella and Solfaterre (also given as "Solfatare"), which were then popular yellow climbers, introduced a few years before. Matthews named the seedling "Augusta" as a "compliment to Mrs. Matthews and her second daughter." The propagation and sale of the rose was given to the nursery of Thorp, Smith, Hanchette & Co., of Syracuse, New York.

Enthusiastic reports about the Augusta rose created intense interest in the horticultural fraternity. Patriotism was also involved. John Feast, the Baltimore nurserymen, praised the rose in the *Florist and Horticultural Journal* (Philadelphia, vol. 2, no. 4, April 1853, pp. 112–13). He encouraged its introduction because of its native origin, and added the stirring command to American horticulturists to "preserve and encourage the raising of all kinds of seedlings, then we shall have plenty without importing trash." Others joined in applauding the Augusta rose for its beauty and fragrance, and its remarkably rapid growth, but there were many who pointed out that it was indistinguishable from the existing variety—Solfaterre. William Henry Prestele diplomatically captioned the plate he made of this rose—"Augusta, or Solfaterre."

The Augusta rose did not long survive its unfortunate debut, notwithstanding its aura of Mount Vernon, its vigorous growth, and its fragrance. By the end of the century it was no longer listed in nursery catalogues.

THE AUGUSTA ROSE.

Plate 73 *Caroline de Sansel Rose*

Caroline de Sansel Rose
(Caroline de Sansal)
Unsigned lithograph; attributed to Joseph Prestele;
engraved on stone, colored; handwritten caption; 27 × 17 cm.
Author's collection.

The evolution of this design is unclear. A plate of the same design shown here, but with the blossom differently colored, has the handwritten caption of "Pius the 9th." Another, with a second bud and also differently colored, is captioned "Barronne Prevost." A third variant, also with two buds but with a tighter arrangement of the leaves, appeared as a nurserymen's plate (28.8 × 22 cm.) and as an illustration in the *Horticulturist*, n.s., vol. 4, no. 11 (November 1854). These last two impressions, including the Prestele handlettered captions, are identical.

The Caroline de Sansal rose belonged to the class of Hybrid Perpetuals (also called "Remontant"). It was introduced to American gardens in the early 1850s and soon became popular. Robert Buist in his *Rose Manual* (Philadelphia, 1851) spoke of it as "quite a rare variety," and praised its large and double blossoms. The *Horticulturist* (1854), in the article accompanying the plate illustrating it, described the blossom as having "a pale silvery blush, with a fleshy tinge in the center." It was one of that small number of roses that remained a favorite, but did it really survive unchanged? In 1910 The Dingee & Conrad Company, West Grove, Pa., the "leading rose growers of America," described the blossom as a "clear, brilliant rose, merging into rosy lilac, tinged with bronze," and not a hint of the earlier "pale silvery blush" which Prestele portrayed.

Caroline de Sansel

Plate 74 *Panache d' Orleans*

Panache d'Orleans
Unsigned lithograph; attributed to Joseph Prestele;
engraved on stone, colored; 32 × 24.5 cm.
From an office copy, *Album of Fruit*, vol. 2, Mount Hope Nursery
(Ellwanger & Barry), Rochester, New York.
Courtesy, Ellwanger and Barry Collection,
Department of Rare Books and Special Collections,
Rush Rhees Library, University of Rochester.

This rose, a Hydrid Remontant (everblooming) was developed by Dauvasse, in 1854, as a sport from the Baronne Prevost. H. B. Ellwanger described it in his *Rose* (1882) as being identical with the parent except that the flowers were striped rosy white. Earlier Francis Parkman said that it was flesh colored and striped with rose and purple. Whatever color its stripes may actually have been, it was not well received by growers and soon disappeared. The fault may have been in the character of its bloom, for Ellwanger said that it wasn't "constant" and soon "ran back to the original." Its name was given both as "Panache d'Orleans" as here, and as "Panachée d'Orleans."

PANACHE D'ORLEANS.

Plate 75 *Persian Yellow Rose*

Persian Yellow Rose
Unsigned lithograph; attributed to Joseph Prestele; engraved on stone, colored; 28.5 × 22.2 cm.
Courtesy, Mrs. Emma Setzer, South Amana, Iowa.

Sir Henry Willock brought this rose from Persia to England in 1837, and within ten years it was being propagated and sold by American nurseries. Francis Parkman, who found relief in gardening from his work as a historian, classed it an Austrian Briar. Andrew Jackson Downing in his *Horticulturist* (May 1848, p. 532) described it as "far superior to the Harrison [Harrison's Yellow] in form and shape of the flower, much more double, and a fine, clear, distinct, yellow in its color. It also blooms very freely." He considered it "truly a charming addition to our collection of roses, large as the latter is. Indeed we should say that the smallest collection is scarcely complete without it." Others spoke of its brilliant yellow color, its profuse bloom, and declared that it cast "all other yellow roses entirely in the shade." It quickly became a general favorite and continued to be sold by nurseries and to be loved by successive generations of gardeners. By 1910 it had become a garden classic and was termed "old fashioned." It is still available from American nurseries.

PERSIAN YELLOW ROSE.

Plate 76 *Spirea*

Spirea
"I Prunifolia fl. pl. II Lanceolata.
I Double Flowering Plum Leaved. II Lance Leaved Spirea."
Unsigned lithograph; attributed to Joseph Prestele;
drawn in chalk on stone, colored; 33 × 24 cm.
Courtesy, Mrs. Joseph Mattes, Amana, Iowa.

I
II
SPIREA
I Prunifolia fl.pl. II Lanceolata.
I. Double flowering Plum leaved II. Lance leaved Spirea.

Plate 77 *Pyrus Japonica*

Pyrus Japonica
Unsigned lithograph; attributed to Joseph Prestele; engraved on stone, colored; 28.5 × 22.2 cm.
Author's collection.

PYRUS JAPONICA.
Shrub Variety

Plate 78 *Double White Flowering Almond Double Crimson Flowering Peach*

Double White Flowering Almond. Double Crimson Flowering Peach
Lithograph; caption (and signature?) trimmed;
attributed to Joseph Prestele; engraved on stone, colored;
originally 28 × 21 cm.(?).
Courtesy, Amana Heritage Society, Museum of Amana History.

Beginning sometime in the 1850s Joseph Prestele created several versions of this subject, as described in the legend accompanying Plate 40. The earlier version shows two vertical and parallel sprays of blossoms. That design was used both as nurserymen's plates and as an illustration in the uncolored and colored editions of the *Horticulturist* (1855). Shown here is what may be the final version with the sprays intertwined. The only examples known to us are in Amana, Iowa, collections, suggesting that Prestele created it after moving there in 1858. The plate reproduced here survived by something of an accident. Many years ago two girls in Amana, Louise and Lisette Krauss, cut out the flowers from a plate and pasted them in their scrapbook along with other richly colored fruit and flower prints.

Apricot Blossoms.

Plate 79 *White Fringe Tree*

Chionanthus Virginica. White Fringe Tree
Unsigned lithograph; attributed to Joseph Prestele; engraved on stone, colored; 31 × 24 cm.
Courtesy, Mrs. Joseph Mattes, Amana, Iowa.

This tree, native to many areas in America from New Jersey, south to Central Florida, and west to western Texas, rarely grows more than thirty feet in height and more often is a shrub sending up several stout spreading stems from a common base.

The tree has long figured in American folklore and folk medicine because of its unusual blossoms and its use in the South for treating yaws. It has also been used as a tonic, a diuretic, and for treating intermittent fever. Its names all allude to its flower: Old Man's Beard, Grandfather Graybeard, Snow-flower and Snow-drop, Flowering Ash and Poison Ash.

Seeds of the tree were sent by Alexander Garden from South Carolina to London in 1757. Three years later Peter Collinson, an English horticulturist, ordered trees from John Bartram, a nurseryman and collector of native plants living near Philadelphia.

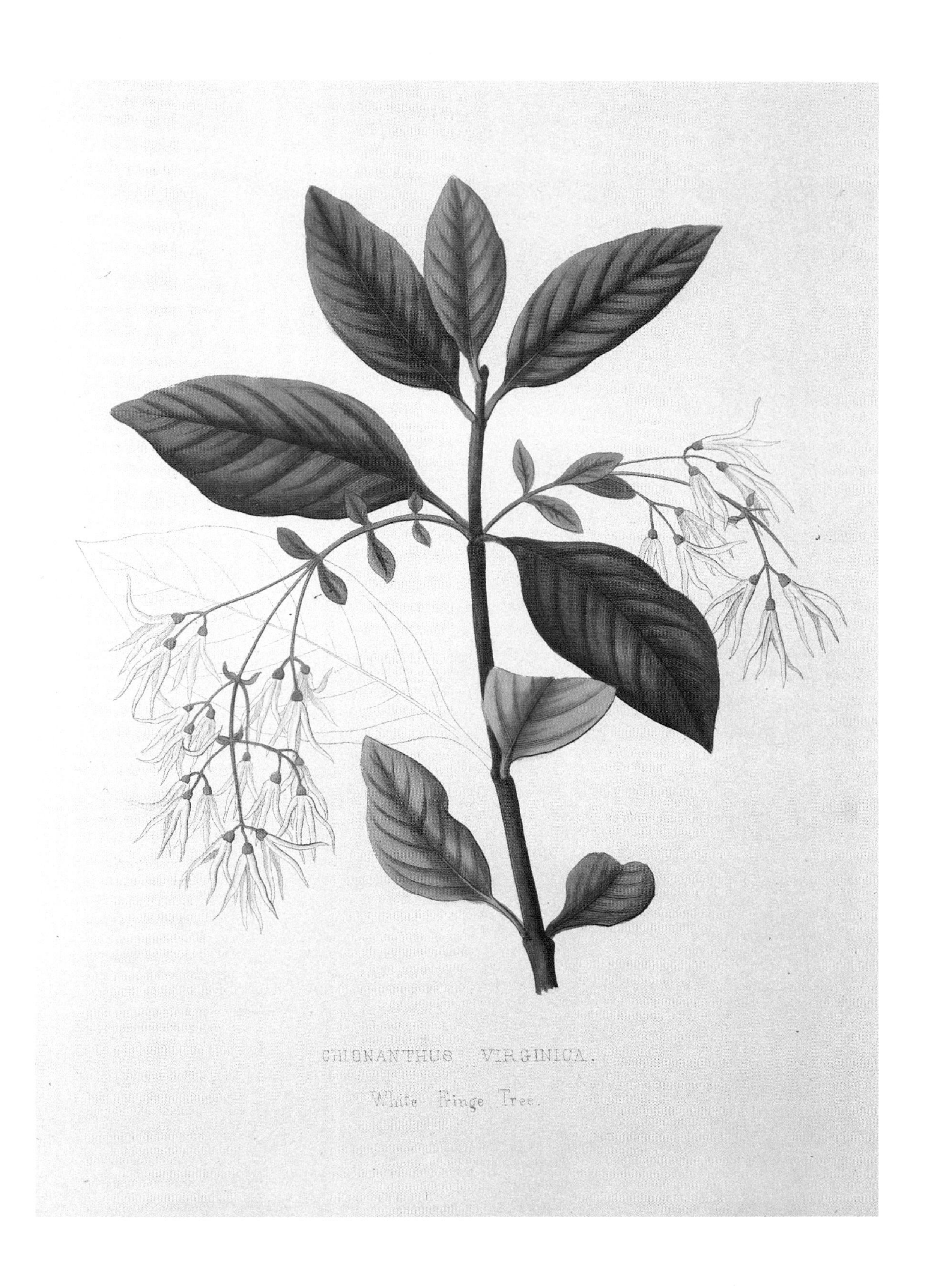
CHIONANTHUS VIRGINICA.
White Fringe Tree.

Plate 80 *Wayfaring Tree*

Viburnum Lantana. Wayfaring Tree
Unsigned lithograph; attributed to Gottlieb Prestele; drawn in chalk on stone, colored; 20 × 17.5 cm.
Courtesy, Mrs. Emma Setzer, South Amana, Iowa.

The Wayfaring Tree is a native of Eurasia and has been long cultivated in southeastern Canada and northeastern United States and occasionally has naturalized.

VIBURNUM LANTANA.
Wayfaring Tree.

Plate 81 *Common European Walnut*

Common European Walnut
(Juglans regia)
Unsigned lithograph; attributed to Joseph Prestele;
engraved on stone, colored; 28.5 × 22 cm.
From a bound assortment of plates issued by the Mount Hope Nursery
(Ellwanger & Barry), Rochester, New York,
to one of their agents in the spring of 1871.
Author's collection.

The many details in this plate, primarily of interest to botanists, are unusual for a nurserymen's plate.

In his *Fruits and Fruit Trees of America* (1845) Andrew Jackson Downing described the European walnut—also referred to as the Persian walnut, of which he said there were a number of varieties—commonly known in America as the Madeira nut. He explained that its green nuts were much used for pickling and that great quantities of the ripe nuts were imported each year. The trees were hardy along the lower Hudson River, and Downing believed that the nut could be profitably grown further south. Trees were in 1845 available in some nurseries.

Since Downing's time the "Common European Walnut" has become known as the English walnut in America and many new varieties are now being grown here.

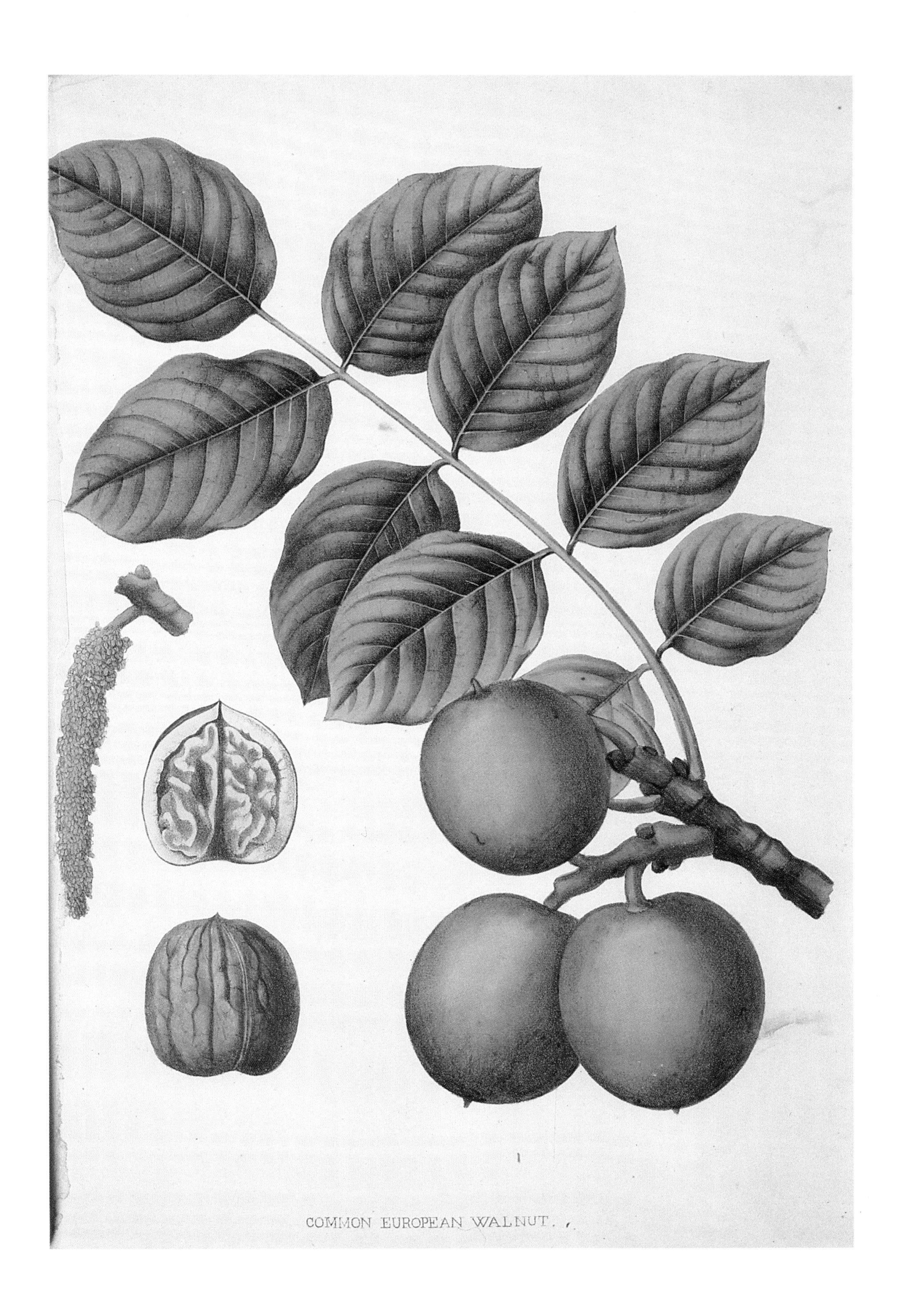

COMMON EUROPEAN WALNUT.

Plate 82 *Downing's Everbearing Mulberry*

Downing's Everbearing Mulberry
Unsigned lithograph; attributed to Gottlieb Prestele; drawn in chalk on stone, colored; 34 × 26.5 cm.
Courtesy, History of Science Collections, Cornell University Libraries, Ithaca, New York.

Charles Downing (1802–1885), pomologist, horticulturist, and author, and the brother of Andrew Jackson Downing, raised this everbearing mulberry from a seed of the *Morus multicaulis* at his Newburgh, New York, nursery about 1846. C. M. Hovey in his *Magazine of Horticulture* (vol. 24, 1858, p. 135), described the tree as vigorous, hardy and productive, with handsome foliage. The tree began bearing when quite young and produced a never-failing crop of—as Hovey put it—the most luscious purplish black fruit measuring from one and a half inches in length and a half-inch in diameter, ripening in succession from July 1 to September 1.

DOWNINGS EVERBEARING MULBERRY

Plate 83 *Trumpet Flower*

Bignonia Radicans. Trumpet Flower
Handwritten addition: "Climbing Shrub var."
(*Campsis radicans.* Trumpetcreeper)
Unsigned lithograph; attributed to Joseph Prestele; engraved on stone, colored; 28 × 22.5 cm.
From a bound assortment of plates issued by the Mount Hope Nursery, (Ellwanger & Barry), Rochester, New York, to one of their agents in the spring of 1871.
Author's collection.

This native vine was common from Pennsylvania to Illinois and southward. Its showy flowers, its long period of summer bloom, and its hardiness and ease of cultivation have long made it a popular garden plant throughout much of America.

Climbing Shrub var.
BIGNONIA RADICANS
Trumpet flower.

Plate 84 *Bouquet Tied with Ribbon*

Bouquet Tied with Ribbon
Unsigned lithograph; attributed to Joseph Prestele; engraved on stone, colored; 30.7 × 25.7 cm.
Courtesy, Mrs. Pauline Schaefer, East Amana, Iowa.

No print answering this description is listed in the Joseph and Gottlieb Prestele catalogues of nurserymen's plates. It may be one of the few designs that Joseph Prestele created as an ornamental print intended for framing. Floral prints were a popular decoration in American homes during the mid-nineteenth century and later, and were published in large numbers by various firms. However, few such prints have the elegance and delicacy of Prestele's bouquet. Examples of this plate before being handcolored are known. (See Fig. 14.)

There is a story that Gottlieb Prestele, when a widower, gave this particular colored plate to a young woman of Amana as an expression of his interest in her, but that she gave him no encouragement.

Plate 85 *Gift Painting; Flowers with Verse*

Gift Painting; Flowers with Verse
Watercolor; attributed to Joseph Prestele;
24 × 15.5 cm., undated (ca. 1858–67).
Courtesy, Mrs. Ferdinand Ruff, South Amana, Iowa.

Violet. Anemone. Crown Imperial.

Garden of my heart now let us see what sweet fruit you bear
And how your flowers grow, which your faithful Jesus seeks
Give Humility Field Violet which fares so well with you
Give Hope the Anemone and Faith the Crown Imperial.

Translation by Elisabeth Kottenhahn.

Veilchen . Annemonen . Keiserkrone .
Herzens-Garten nun laß sehen, Was du trägst für süße Frucht,
Und wie deine Blumen stehen, Die dein treuer Jesus sucht.
Gieb die Hoffnungs Annemonen, Und des Glaubens Kaiserkron.

Plate 86 *Blue Vase with Strawberries*

Blue Vase with Strawberries
Unsigned watercolor, by Joseph Prestele; 50 × 39 cm.
Courtesy, Mrs. Ferdinand Ruff, South Amana, Iowa.

Joseph Prestele and his son Gottlieb made many decorative watercolors, often to give as gifts for friends or to commemorate some special event such as a wedding. The father seems to have been intrigued by the grace and color of strawberry plants when in fruit, as witness the number of airy compositions of these he made as nurserymen's plates and magazine illustrations. Here he created an improbable arrangement, for it is unlikely that strawberry vines in bearing could have been arranged in such a manner, but he did make a graceful design, enhanced by the blue of the vase.

Plate 87 *Guide to Meditation*

Guide to Meditation
Watercolor, by Gottlieb Prestele
(signed: "Gemalt von G. Prestele"); 42 × 32 cm.
Courtesy, Elmer P. Graesser, West Amana, Iowa.

This meditation on the omnipotence of God has an intricate pattern of religious symbols with minute and carefully painted details. The work resembles German folk art. Few other examples of Gottlieb's skill as a creative artist are known.

Below "Seelen Spiegel" [Soul Mirror] in the center of the painting are found an eye, ear, rod, and other reminders that God sees, hears, and rules all. Then comes:

Whoever contemplates this in his heart
Will receive many good things.
Whoever forgets this wisdom
Is more a heathen than a Christian.
Therefore remember God's presence
Who reveals himself to the Believers.

And the final reminder:

Your God is everywhere
And can easily find you.
He can see into your heart.
Oh, beware of sins.

Translation by Elisabeth Kottenhahn.

IHR ALTEN UND IHR
JUNGEN SEHT,
WAS HIER VERDECKT
GESCHRIEBEN STEHT.
SEELEN SPIEGEL.
Wer dieß im Herzen oft bedenkt
Dem wird viel Guts dadurch geschenkt
Wer diese Wahrheit gern vergißt
Ist mehr ein Heide als ein Christ
Drum denk an Gottes Gegenwart
Die sich den Frommen offenbart.
Dein Gott ist überall;
Und weiß dich wohl zu finden,
Er sieht auch in dein Herz,
O hüte dich vor Sünden!

Plate 88 *Lady's Slipper*

Cypredium Spectabile. Lady's Slipper
(Frauenschuh)
(*Cypripedium reginae* Walt. Showy Ladyslipper.)
"Gez[eichnet] und Gemalt von G. Prestele. Amana, Iowa."
Watercolor; 53.5 × 40.5 cm.
Courtesy, Mrs. Lena Unglenk, Amana, Iowa.

Cypripedium spectabile.
Ladys Slipper

Plate 89 *Showcard*

Showcard
Lithograph, by William Henry Prestele;
chalk on stone, colored; copyrighted 1872;
72 × 57.6 cm. (sheet size).
Courtesy, Prints and Photographs Division,
Library of Congress.

While living in Bloomington, Illinois, William Henry Prestele produced this showcard in 1872 for use by shopkeepers and nurserymen who then inserted their names and commercial messages in the open center of the piece. Perhaps because of its large size Prestele had it printed by C. Hamilton & Co., St. Louis.

A copy of this showcard was used by Prestele in 1884 when preparing a tribute to Charles Downing, the Newburgh, New York, horticulturist and author who had befriended him during some crisis in his life. Prestele wrote his appreciation in the center of the print (quoted on p. 107). The Downing tribute is in Crawford House, the headquarters of the Historical Society of Newburgh Bay and the Highlands, New York.

Original. W. H. Prestele
C. HAMILTON & Co. LITH. ST LOUIS, MO
Published by Wm H. PRESTELE, Bloomington Ills
SHOW CARD.

Plate 90 *Flemish Beauty*

Flemish Beauty
"(Sample Plate.)"
"Lith. & col[d]. by W. H. Prestele, Artist, Iowa City, Ia."
Lithograph; drawn on stone, colored; ca. 1875; 23.8 × 15 cm.
Frontispiece from W. H. Prestele's *187[9]. Catalogue of Fruit and Flower Plates Drawn from Nature. Lithographed and Colored by Wm. H. Prestele, Son of the late Joseph Prestele, Sen., of Amana Society, Homestead, Iowa. Iowa City, Johnson Co., Iowa.* Republican Job Print, Iowa City, Iowa.
Author's collection.

The Flemish Beauty pear originated in Belgium. It was being grown in the Boston area beginning about the mid-1830s and was known there as the "Barnard pear," one of the many names it acquired. Its huge size, its "extremely melting and luscious" flavor, as C. M. Hovey described it, and other good qualities made it a favorite for many years among home growers. But it was not of commercial value because the fruit required special handling; its popularity waned with the passing of the home orchard.

FLEMISH BEAUTY.
(SAMPLE PLATE)
Lith. & Col^d. by W.H.PRESTELE, Artist. Iowa City, Ia.
PUB^R OF COL^D LITHOGRAPHS OF FRUITS & FLOWERS, SHOW
CARDS, AND ALL SORTS OF COMMERCIAL WORK, SUCH AS
LETTER HEADS, BILL HEADS, DRAFTS CHECKS, LABELS,
ENVELOPES &.&.

Plate 91 *Rose of Sharon*

Althea or *Rose of Sharon*
"Lith. & Col^{d}. by W. H. Prestele, Iowa City, Iowa."
Lithograph; drawn in chalk on stone, colored; 18.5 × 11.5 cm.
Courtesy, Mrs. Emma Setzer, South Amana, Iowa.

ALTHEA or ROSE of SHARON
Lith & Cold by W.H. Preston Iowa City Iowa.

Plate 92 *Seven Sisters*

Seven Sisters
"Lith. & Col[d]. by W. H. Prestele. Iowa City, Ia."
Lithograph; drawn in chalk on stone, colored; 29.5 × 23 cm.
Courtesy, Mrs. Emma Setzer, South Amana, Iowa.

The Seven Sisters rose, so named because of the number of blossoms in the clusters, was a multiflora whose blush-colored petals were tinged and striped with various shades of pink and red. It was a tender variety, scorned by rose fanciers. H. B. Ellwanger of Rochester, New York, who was ordinarily restrained in his published comments, could not withhold his contempt for it in his book *The Rose* (1882). Tree peddlers made large sales of it, he said, "by means of exaggerated colored plates, accompanied by untruthful descriptions." One suspects he had seen the extraordinary effusion reproduced here.

SEVEN SISTERS.
Lith. & Col^d by W.H.PRESTELE. Iowa City. Ia.

Plate 93 *Mock Orange*

Syringa or *Mock Orange*
"Lith. & Col^{d}. by W. H. Prestele. Iowa City, Iowa."
Lithograph; drawn in chalk on stone, colored; 18.5 × 11.5 cm.
Courtesy, Mrs. Emma Setzer, South Amana, Iowa.

SYRINGA or MOCK ORANGE.
Lith & Col.d by W.H. Prestele. Iowa City. Iowa.

Plate 94 *Dielytra Spectabilis*

Dielytra Spectabilis
(*Dicentra spectabilis*. Common Bleedingheart.)
"Lith. & Col^{d}. by W. H. Prestele, Iowa City, Iowa."
Lithograph; drawn in chalk on stone, colored; 19 × 13 cm.
Courtesy, Mrs. Emma Setzer, South Amana, Iowa.

Robert Fortune, the plant collector, brought the *Dicentra spectabilis* from China to England in 1846. It first flowered in America in 1851 where, because of its ease of propagation, its beauty, and, one suspects, its novel and sentimental form, it soon became a familiar and much-beloved garden plant. It acquired its popular name in England to become another of several flowers called "Bleeding Heart."

The example shown here is one of the nurserymen's plates made at Iowa City by William Henry Prestele after moving there about the mid-1870s.

DIELYTRA SPECTABILIS
Lith. & Cold. by W.H. Prestele. Iowa City. Ia.

Plate 95 *Vitis Lincecumii*

Vitis Lincecumii var. glauca
Unsigned watercolor; painted by William Henry Prestele; 43 × 35 cm.
Courtesy, U. S. Department of Agriculture,
Beltsville Agricultural Research Center, and
J. R. McGrew, plant pathologist.

In 1887, Thomas Volney Munson (1843–1913), who had won recognition for his work in developing many new varieties of grapes, was appointed special agent by the Pomological Division of the United States Department of Agriculture, to prepare a monograph on the native species of grapes. The project necessitated Munson traveling many thousand miles throughout America to locate, collect, and identify specimens of native grapes. William Henry Prestele, appointed as staff artist the same year as Munson, was assigned to prepare the plates from dried specimens, photographs, and sketches supplied by Munson. The artist made color sketches of the separate parts of the total design and sent them to Munson at Denison, Texas, for review. When all the preliminary sketches had been approved, Prestele completed the final plate. The artist spent about six years at that task during which he prepared plates of some thirty species and subspecies. The collection is preserved at the department's Beltsville Agricultural Research Center in Maryland.

Munson's report—completed by 1891—was never published by the department as had been intended, and all but a fragment of the manuscript has been lost. Fortunately, the material was included in Munson's best-known work, *Foundations of American Grape Culture* (1909).

(J. R. McGrew, plant pathologist, USDA, "The W. H. Prestele Paintings of the American Species of Grapes," *T. V. Munson Memorial Vineyard Report*, vol. 1, no. 2, April 1, 1981. Grayson County College, Denison, Texas.)

Glossary

Technical Terms Used by Joseph Prestele and Isaac Sprague in Their Signatures on Plates

AD. NAT. DEL.	"ad naturam delineavit," as in "Sprague ad nat. del."; "Sprague drew from nature."
DEL.	"delineavit," as in "Sprague del."; "Sprague drew."
IN LAP. SC. ET LITH.	"in lapis sculpsit et (lithographed?)," as in "Prestele in lap. sc. et lith."; "Prestele engraved on stone and printed." ("Lithograph, lithographed" was used either by the person who drew the image on the stone, or by the lithographic printer.)
OMNES SC.	"omnes sculpsit," as in "Prestele omnes sc."; "all engraved by Prestele." Used on the first plate of a series in a book when the name of the engraver is not repeated on each plate.
OMNES IN LAP. SC.	"omnes in lapis sculpsit," as in "Prestele omnes in lap. sc."; "all engraved on stone by Prestele." Used in the same manner as "omnes sc."
PINX.	"pinxit," "painted."
SC.	"sculpsit," as in "J. Prestele sc."; "Joseph Prestele engraved."

Appendix A

Annotated Checklist of Published Plates for *The Forest Trees of North America*, by Asa Gray, Together with Preliminary Watercolors and Proofs

The published plates (twenty-three all told) are on unwatermarked rag paper. The sheets are 34 × 27 cm., except for two folding plates (number 3 and 4) which are 34.8 × 45.5 cm. Unless otherwise noted the plates are printed in black ink. All plates are tinted with polychrome colors.

Prestele was kept busy with making these plates from 1849 to 1851, and during those years he seems to have worked on all of them. With some, he engraved, printed, and colored the plates whose designs had been drawn by Isaac Sprague. With others, his contribution was limited to coloring. At times he had difficulty keeping his creative and artistic urges under control; his offers to make some of the preliminary sketches were ignored by Asa Gray, who found it convenient to work with Isaac Sprague who lived nearby in Cambridge. When he took it upon himself to make slight alterations in Sprague's designs for artistic or scientific reasons, he usually met with disapproval from Gray. Many of the plates are unsigned but can be attributed to Prestele. While his letters to Gray and others provide many clues as to the plates he produced, or aided in producing, those brief references are too incomplete to provide a coherent account of his contribution.

The watercolor and proofs noted below are in the collections of the Hunt Institute for Botanical Documentation, Pittsburgh, Pennsylvania (hitherto identified as HIBD), and the Ellwanger and Barry Collection, Department of Rare Books and Special Collections, Rush Rhees Library, The University of Rochester (identified as RU-E&B). The letters quoted, from Joseph Prestele to Asa Gray, and to Isaac Sprague are in the Gray Collection, Gray Herbarium of Harvard University. Those to John Torrey are from the Torrey Collection at the New York Botanical Garden. Letters to Prestele from officers of the Smithsonian Institution are from the Smithsonian's Archives. All are quoted with permission.

"Plate 1. MAGNOLIA GRANDIFLORA Large Flowered Magnolia"
"Sprague del. Sonrel in lap. Tappan & Bradford's lith."

Drawn in "chalk" (a greasy crayon) on stone; printed in green ink; colored by Joseph Prestele. a. Proof. Printed in green ink; uncolored. Text as above. (HIBD). b. Proof. Printed in black, before letters; uncolored. (HIBD). c. Proof. Colored plate, no text. Written across the top is "Ruh___12c___" (perhaps a colorist's plate). (HIBD).

"Plate 2. MAGNOLIA GLAUCA Small Magnolia"
"Sprague ad nat. del."

Drawn in "chalk" on stone; printed in green ink by Tappan & Bradford, Boston (Prestele to Gray, February 13, 1865); colored by Prestele. a. Proof. Printed in black; uncolored. Text as above. (HIBD).

"Plate 3. UMBRELLA MAGNOLIA Umbrella Tree"
"Sprague del. Sonrel in lap. Tappan & Bradford's lith."

Folding sheet. Drawn in "chalk" on stone; printed in green ink except for a portion of the lower right-hand section which was printed in red. Colored by Joseph Prestele who explained to Isaac Sprague that this was a very difficult plate to color, particularly the fruit (Prestele to Sprague, January 22, 1851). Prestele increased his charge for coloring the plate to 10 cents each (Prestele to Sprague, February 11, 1851). On the latter date he sent 390 plates that he had colored to Sprague, Gray then being in Europe. a. Proof. Printed in black, before letters; uncolored. (HIBD). b. Proof. Printed in green, before letters; uncolored. (HIBD).

"Plate 4. MAGNOLIA AURICULATA Ear-lobed Umbrella Tree"
"Sprague del. Sonrel in lap. Tappan & Bradford's lith."

Folding sheet. Drawn in "chalk" on stone; printed in green ink. Colored by Joseph Prestele who sent

Isaac Sprague 250 finished plates on December 2, 1850, and 139 more on February 11, 1851 (Prestele to Sprague). a. Watercolor. Unsigned, with penciled legend "MAGNOLIA AURICULATA. Ear-lobed Umbrella Tree." (HIBD).

"Plate 8. LIRIODENDRON TULIPIFERA Tulip Tree"
"Sprague ad nat del."

Engraved on stone (by Joseph Prestele?), who colored the plate. Printed by Tappan & Bradford (Prestele to Gray, February 13, 1865). a. Proof. Printed in black; text as above. Signed on back "Oct. 16, 1852." (HIBD)

"Plate 10. TILIA AMERICANA American Linden"
(No signature)

Engraved on stone by Joseph Prestele; printed and colored by him. While Gray was in Europe, Prestele thought to improve Sprague's drawing by a slight alteration (Prestele to Sprague, January 22, 1851). When Gray returned he rejected the version Prestele had engraved and colored. Prestele offered to redo it. The final version that appears to have been accepted seems to be the original one.

"Plate 20. ACER RUBRUM Red Maple."
(No signature)

Engraved on stone; printed and colored by Joseph Prestele. When Sprague was critical of Prestele's colored proof (Gray being in Europe), the artist replied that he could improve the plate by making the "flowers more light." He added that he had increased his charge for coloring this plate to 10 cents each (Prestele to Sprague, February 11, 1851).

"Plate 25. ACER SPICATUM Mountain Maple."
(No signature)

Engraved on stone, printed and colored by Joseph Prestele. Prestele wrote Sprague that he would "make the shades in the fruit behind the leaf more light" (February 11, 1851). a. Proof. Caption crudely lettered; colored; lacks "Plate" and plate number. (RU-E&B)

"Plate 27. AESCULUS GLABRA Ohio Buckeye."
(No signature)

Engraved on stone; printed and colored by Joseph Prestele. Prestele explained to Dr. S. F. Baird, the assistant secretary of the Smithsonian Institution (January 27, 1852) that coloring this plate was a very slow and very exacting task. "I can do very few a day."

"Plate 30. AESCULUS DISCOLOR"
"Sprague del. Prestele sc. et lith."

Engraved on stone and colored by Joseph Prestele. Prestele wrote Sprague that this plate was particularly difficult to color, and that he had increased his charge to 12 cents a piece. "This enables me to do them so much better and with more care." (Prestele to Sprague, January 22, 1851; February 11, 1851). The artist sent Sprague 100 copies of the finished plate on December 2, 1850, and 175 more on February 11, 1851.

"Plate 31. AESCULUS PARVIFLORA"
"Sprague del. Prestele sc. et lith."

Engraved on stone and colored by Joseph Prestele. The artist sent Sprague 100 copies of the finished plate on December 2, 1850, and 175 more on February 11, 1851 (Prestele to Sprague, December 2, 1850; February 11, 1851). a. Proof. Colored; caption crudely lettered. Lacks "Plate" and plate number. (RU-E&B)

"Plate 34. ROBINIA PSEUDACACIA Common Locust Tree"
"I. Sprague del. J. Prestele sc."

Printed and colored by Joseph Prestele.

"Plate 35. ROBINIA VISCOSA Clammy Locust Tree"
(No signature)

(Drawn by Sprague?) Engraved, printed, and colored by Joseph Prestele (Prestele to Gray, February 13, 1865). Prestele sent Sprague 350 copies of the finished plate on December 2, 1850, and 50 more on February 11, 1851 (Prestele to Sprague). a. Proof. Printed in black on India paper, before being lettered. (HIBD). b. Proof. Printed in black, before lettered. (HIBD). c. Proof. Colored; caption crudely lettered; lacks "Plate" and plate number. (RU-E&B)

"Plate 39. CERCIS CANADENSIS Red Bud."
(No signature)

(Drawn by Sprague?) Engraved, printed, and colored by Prestele. The artist increased his charge for coloring this plate to 10 cents each (Prestele to Sprague, February 11, 1851). He sent Sprague 150 copies of the finished plate on December 2, 1850, and 100 more on February 11, 1851 (Prestele to Sprague).

"Plate 40. GYMNOCLADUS CANADENSIS
Kentucky Coffee Tree"
"J. Prestele sc. I. Sprague del."

Printed and colored by Joseph Prestele. Prestele wrote Sprague on February 11, 1851, that he had engraved this plate again upon Sprague's recommendation. "My charge will be only $2 although it was very difficult to engrave, particularly the fruit" (Prestele to Sprague).

"Plate 41. GLEDITSCHIA TRIACANTHOS Honey Locust."
"Sprague del."

Engraved on stone. Engraved, printed, and colored by Joseph Prestele. Prestele had difficulty in making a plate that Gray would approve. While Gray was in Europe, Prestele's printer struck off one hundred plates which Prestele's children began coloring (Prestele to Gray, February 24, 1851). When Gray returned and objected to the plate, Prestele said he would alter it without charge if

Gray would pay $8 for the coloring of the 100 plates, and send him a supply of paper on which to print the new ones (Prestele to Gray, April 1, 1852). Gray sent the check and returned the proofs of the corrected version with a few further changes which Prestele then made. He said he would take special pains in coloring the new ones (Prestele to Gray, April 21, 1852). a. Proof. Caption crudely lettered; colored. Lacks "Plate" and plate number. (RU-E&B)

"Plate 46. PRUNUS AMERICANA Wild Plum"
(No signature)

(Drawn by Sprague?) Engraved on stone, printed, and colored by Joseph Prestele. On February 11, 1851, Prestele wrote that he was increasing his charge for coloring this plate to 10 cents each (Prestele to Sprague). He sent Sprague 100 copies of the finished plate on December 2, 1850, and 100 more on February 11, 1851 (Prestele to Sprague).

"Plate 47. PRUNUS CHICASA Chickasaw Plum"
"Sprague del. Prestele in lap sc. et lith."

Engraved on stone; colored by Joseph Prestele. Prestele sent Sprague 150 copies of the finished plate on December 2, 1850 (Prestele to Sprague). a. Proof. Printed in black with text, as above. (HIBD)

"Plate 48. CERASUS PENNSYLVANICA Wild Red Cherry"
(No signature)

(Drawn by Sprague?) Engraved, printed, and colored by Joseph Prestele.

"Plate 49. CERASUS VIRGINIANA Choke Cherry"
(Drawn by Sprague?) Engraved, printed, and colored by Joseph Prestele (Prestele to Gray, February 13, 1865). Prestele sent Sprague 300 copies of the finished plates on December 2, 1850, and 100 more on February 11, 1851 (Prestele to Sprague). a. Proof. Printed in black on India paper before lettered. (HIBD)

"Plate 50. CERASUS SEROTINA Wild Black Cherry"
"Sprague del. Prestele in lap sc. et lith."

Engraved on stone; colored by Joseph Prestele. A colored proof of (possibly) this plate was sent to Gray on December 19, 1849 (Prestele to Gray). A few months later he asked Gray to send him paper so he could begin printing and coloring some of the plates, proofs of which he had sent him," viz. of the . . . Cerasus." He sent Sprague 150 copies of the *Cerasus Serotina* on December 2, 1850, and 100 more on February 11, 1851 (Prestele to Sprague). a. Proof. Printed in black; text as above. (HIBD)

"Plate 52. PYRUS CORONARIA American Crab Apple"
"I. Sprague del. J. Prestele sc."

Printed and colored by Joseph Prestele.

"Plate 63. CORNUS ALTERNIFOLIA Dogwood."
"Sprague del."

Engraved on stone, printed, and colored by Joseph Prestele.

Appendix B

Partial Checklist of U.S. Army Reports of Western Expeditions Containing Illustrations Produced Wholly or in Part by Joseph Prestele

The Joseph Prestele illustrations have been identified by signatures on the plates, by credits in the accompanying text, or by references in Prestele's letters to John Torrey and Asa Gray. It can be assumed that additional examples await identification.

I. *Pacific Railroad Surveys*

The Pacific Railroad Surveys were reprinted in *Reports of Explorations and Surveys to Ascertain the Most Practicable and Economical Route for a Railroad from the Mississippi River to the Pacific Ocean. Made under the Direction of the Secretary of War, in 1853–(56).* 12 vols. in 13 books. 33rd Congress, 2nd Session. House of Representatives, Exec. Doc. No. 91. (Washington: 1855–60.)

Lieutenant Edward Griffin Beckwith, *Report of Explorations for a Route for the Pacific Railroad, on the Line of the Forty-First Parallel of North Latitude, 1854* (33rd Congress, 2nd Session. Senate, Exec. Doc. No. 78. Volume II, part 2, 1854–55) contains "Report on the Botany of the Expedition," by John Torrey and Asa Gray, pp. 119–32, with ten uncolored plates, engraved by Prestele and printed by Ackerman. No artist credits are on these plates, but Andrew Rodgers (*John Torrey*, p. 237) states they were engraved by Prestele. Torrey sent the artist proofs of Beckwith's plates in April 1857. Beckwith's *Report* includes information collected on the initial phase of the expedition while under the command of Captain J. W. Gunnison, who was killed by Indians in Utah and whom Beckwith succeeded as commanding officer.

Brevet Captain John Pope, *Report of Exploration of a Route for the Pacific Railroad, near the Thirty-Second Parallel of North Latitude, from the Red River to the Rio Grande, 1854*. The *Report* was published in a small edition in 1855 and reprinted in Volume II of the formal edition of the Pacific Railroad Surveys. The latter edition includes a botanical report of nineteen pages by Torrey and Gray (Volume II, part 2, pp. 159–78), and ten uncolored botanical plates with the following credits: "Sprague del. P.S. Duval & Co., Lith. Philada. Prestele sc."

Lieutenant Amiel Weeks Whipple, *Explorations and Surveys for a Railroad Route . . . Near the Thirty-Fifth Parallel . . . in 1853–54*. Volume IV (1856), part 5, no. 3, pp. [27]–58, "Description of the Cactaceae," by George Engelmann, M.D., of St. Louis, and J. M. Bigelow, M.D., has twenty-four uncolored plates, drawn by Paulus Roetter of St. Louis and by H.B. Mollhausen, the staff artist. They were printed by Ackerman. These appear to have been engraved by Prestele as were the twenty-five plates in the "Description of the General Botanical Collection," by John Torrey (No.4, pp. 61–182). The plates are unsigned but Torrey in his introductory note (p. 59), dated January 12, 1857, says, "All the engravings have been done on stone by Prestele, who excels in this branch of the art."

Lieutenant R. S. Williamson, *Report of Explorations for the Pacific Railroad Upon Routes in California Connecting with Routes Near the Thirty-Fifth and Thirty-Second Parallels*, 1853–54, and published as Volume V of the Pacific Railroad Survey *Reports*. Part 3, pp. 5–15, "Botanical Report by E. Durand and T. C. Hilgard, M.D.," has eighteen plates printed by Ackerman, some of which may have been engraved by Prestele. Article VII, pp. 359–70, Torrey's "Description of the Plants Collected Along the Route," has ten plates which Torrey stated (p.359) that Prestele engraved, and which Ackerman printed.

II. *Mexican Boundary Survey*

Major William Helmsley Emory, *Report on the United States and Mexican Boundary Survey*. 34th Congress, 1st Session, Senate Exec. Doc. No. 3. 3 vols (Washington: 1857–59). Volume II, part 1 (1858), "Botany of the Boundary," by John Torrey (pp. [29]–236), has sixty-one plates, some engraved by Prestele. On May 15, 1856, the artist wrote Torrey that he had received authorization from Washington to engrave botanical plates at $10 each. February 4, 1857, he sent four proofs of cacti to Torrey, commenting on his difficulties with Plates XVII and XVIII. April 27, 1857, he acknowledged receipt of "2 drawings for Major Emory's Plants" and said he would send four proofs of Emory's plants and the stone to Ackerman the following week. On January 14, 1858, Prestele reported problems with drawing and engraving some of the pines, possibly those published as Plates 53–59, Volume II, part 1, of Torrey's report.

III. *Exploration of the Valley of the Great Salt Lake*

Captain Howard Stansbury, *Exploration and Survey of the Valley of the Great Salt Lake of Utah.* 32d. Congress, Senate, Special Session, March 1851, Exec. Doc. No. 3. (Philadelphia: Lippincott, Grambo & Co., 1852). This work was later reprinted several times.

Appendix D, pp. [381]–97, "Catalogue of Plants Collected by the Expedition," by John Torrey, has nine plates, engraved (but not signed) by Prestele, and printed by Ackerman. Prestele reported starting work on these plates in a letter to Torrey on September 26, 1851, adding that the drawings are "thickly placed with figures" and that he would make four "drawings" (i.e., engravings) on one stone. On November 24, 1851, he wrote Torrey that he had finished his work, including reengraving four drawings on a stone that had broken. These subjects he gives as "Huechera rubescens, Perityle suffruticosa, Amblirion pudicum, and Chenatis densifolia." Some of these names were changed in preparing the captions for the published plates.

Appendix C

Checklist of Nurserymen's Plates Signed by, or Attributed to, Joseph Prestele, Sr., and His Sons Gottlieb and William Henry

All plates listed are hand-colored lithographs except those which are uncolored or which may be watercolors, and are so noted. No attempt has been made to identify the specific lithographic techniques used, but they include engravings on stone, images drawn in "chalk" on stone, possibly transfers drawn in "tusche" (lithographer's ink), and other variations of the lithographic technique. Examples of many plates noted as having been located were reported to the author but not actually seen by him, hence it has not always been possible to mention variations in the designs.

Joseph Prestele, Sr., and his son Gottlieb signed relatively few of their plates; more of their work carries the Amana Society's imprint. A considerable number were published anonymously. Some were sold to nurseries and publishers who added their own imprint. Unsigned plates engraved on stone can be attributed to Joseph, Sr., on the basis of style, letter design, references in various contemporary sources, and his use of an engraving technique not employed by his sons. Fortunately, William Henry appears to have signed all of his published work.

The entries in the checklist include titles from four catalogues (three are actually broadside lists) of plates by the Presteles. Many examples have been located. As others turn up, tentative assumptions about them can be corroborated or corrected.

In addition to serving as a listing of the nurserymen's plates made by the Presteles, it is also a list of the plants widely grown in America during the middle and later years of the nineteenth century, which students of American gardening history should find of interest. The varieties chosen for illustration were popular and saleable, or were new varieties that nurserymen wished to promote. The present scientific names of plates are given where the original names might be confusing.

The nomenclature used for the headings in the Apple and Pear sections of the checklist is based upon two publications issued by the U. S. Department of Agriculture, Bureau of Plant Industry, compiled by W. H. Ragan: *Nomenclature of the Apple*, Bulletin No. 56, first issued in January 1905 (Washington: 1926); and *Nomenclature of the Pear*, Bulletin No. 126, first issued in June 1908 (Washington: 1926).

SOURCES

Where many copies of individual plates have been found, the location of only a few are given, usually in public collections.

AAS — Mrs. Adolph Schmieder, Middle Amana, Iowa.

AAS-3 — Mrs. Adolph Schmieder, *Catalogue*, ca. 1875.

AEG — Elmer P. Graesser, West Amana, Iowa.

AES — Mrs. Emma Setzer, South Amana, Iowa.

AJM — Mrs. Joe Mattes, Amana, Iowa.

AMAH — Amana Heritage Society, Museum of Amana History, Amana, Iowa.

APC — Private collection.

APS — Mrs. Pauline Schaefer, East Amana, Iowa.

AS — Amana Society, Amana, Iowa.

CU — History of Science Collections, Cornell University Libraries, Ithaca, New York.

CU-1. A collection of unbound and untrimmed plates by Joseph, Sr., and/or Gottlieb Prestele, 34 × 26.5 cm.

CU-2. *List of Fruit-and Flower-Plates. Drawn from Nature. Lithographed and Colored by Joseph Prestele, Sen., Amana Society, Homestead P. O., Iowa Co., Iowa.* (1 page; ca. 1860.)

CUM — Reference Division, Albert R. Mann Library, New York State College of

Agriculture and Life Sciences, at Cornell University, Ithaca, New York.

CUM-1. Bound collection of plates by Joseph, Sr., and/or Gottlieb Prestele. Untitled; no date, plates trimmed to 28.5 × 22.5 cm.

CUM-2. *Catalogue of Fruit-and Flower-Plates. Drawn from Nature. Lithographed and Colored by Amana Society, Homestead P. O., Iowa.* (1 page; c.1875. Bound with CUM-1.)

CvR Author's collection.

CvR-1. Untitled bound collection of nurserymen's plates by various publishers, c.1860. Plates trimmed to various sizes, the majority 28.2 × 22 cm.

CvR-2. *Album of Fruit.* Bound collection of nurserymen's plates by various publishers. Issued by Ellwanger & Barry to their agent, F. W. Kelsey, in the spring of 1871. Plates trimmed to 28.5 × 22 cm.

CvR-3. D. M. Dewey, *The Colored Fruit Book, for the Use of Nurserymen.* (Rochester, N.Y., 1859.) Plates trimmed to 28.5 × 22 cm.

CvR-4. (William Henry Prestele.) *Catalogue of Fruit and Flower Plates Drawn from Nature, Lithographed and Colored by Wm. H. Prestele, Son of the Late Joseph Prestele, Sen., of Amana Society, Homestead, Iowa. Iowa City, Johnson Co., Iowa.* (Iowa City: Republican Job Print, 1879.)

CvR-5. (Bloomington Nursery, Bloomington, Ill.) Plate book assembled and bound after 1876; plates by various publishers, most with the imprint of W. K. Phoenix, Bloomington Nursery, trimmed to 27 × 21 cm. Untitled.

OSU Department of Special Collections, University Libraries, Ohio State University, Columbus, Ohio.

Specimen Book of Fruits & Flowers, Carefully Drawn from Nature. Lithographed and Colored by G. Prestele, Amana Society, Iowa County, Iowa. No date; plates trimmed to 29 × 23.2 cm.

RU Ellwanger & Barry Collection. Department of Rare Books, Manuscripts and Archives, University of Rochester, Rochester, New York.

RU-1. *Album of Fruit, No. 2.* No date; plates trimmed to 32 × 24.5 cm.

RU-2. Unbound plates; slight variations in size, about 34 × 27.7 cm.

LIST OF FRUIT- AND FLOWER-PLATES. DRAWN FROM NATURE.

LITHOGRAPHED AND COLORED BY JOSEPH PRESTELE Sen., AMANA SOCIETY, HOMESTEAD P. O., IOWA CO., IOWA.

PRICE OF PLATES FROM 25 AND UPWARDS.

List of Fruit-and Flower-Plates. Drawn from Nature. Lithographed and Colored by Joseph Prestele Sen., Amana Society, Homestead P. O., Iowa Co., Iowa, ca. 1860. Courtesy, History of Science Collections, Cornell University Libraries, Ithaca, New York.

Although this *List* states that the plates on it were made by Joseph Prestele, actually it includes plates that Gottlieb made or on which he did a considerable part of the work. It is the earliest such list that has been located, and it can be dated between 1858 when the Presteles arrived at Homestead, Iowa, and 1867 when Joseph died. The different categories of plant material shown on the *List*, and the number of subjects included in each of them are an indication of what plants were popular at the time, at least among the nurseries using Prestele plates. The number of apple plates was almost double those of pears, the second most frequently planted fruit. Plums were more popular than peaches; strawberries and grapes tied with thirteen each. Among the ornamental plants, roses were the most sought after, sixteen varieties being included. The total number of plates being sold at that time by the Presteles was 233.

Catalogue

OF

FRUIT- AND FLOWER- PLATES.

DRAWN FROM NATURE.

LITHOGRAPHED AND COLORED BY AMANA SOCIETY, HOMESTEAD P. O., IOWA.

Apples.

1 Alexander,
2 Astrachan Red,
3 Autumn Strawberry,
4 Bailey's Sweet.
5 Baldwin,
6 Belmont,
7 Ben Davis (New York Pippin.)
8 Benoni,
9 Black Apple
10 Canada Red,
11 Cooper,
12 Domine,
13 Duchesse of Oldenburg.
14 Dutch Mignonne,
15 Eearly Harvest,
16 Early Joe,
17 Early Strawberry,
18 Esopus Spitzenburg,
19 Fameuse,
20 Fall Wine
21 Fulton
22 Genesee Chief,
23 Golden Russet,
24 Gravenstein,
25 Grimes Golden
26 { Hubbardston Nonsuch, Michael Henry Pippin, } on one plate,
27 Jersey Sweet,
28 Jonathan,
29 Keswick Codlin,
30 King of Tompkins Co.,
31 Lady Apple,
32 Maidens Blush,
33 Minister,
34 Monmouth Pippin,
35 Mother,
36 Newtown Pippin,
37 Newtown Spitzenburg,
38 Northern Spy,
39 { Pecks Pleasant, White Bellflower, } on one plate,
40 Perry's Russet.
41 Pewaukee
42 Pomme Grise,
43 Porter,
44 Rambo,
45 Raule's Jannet,
46 Red Detroit,
47 Reinette Canada,
48 Rhode Island Greening,
49 Ribston Pippin,
50 Rome Beauty,
51 Roman Stem,
52 Roxbury Russet,
53 Seek-no-further,
54 St. Lawrence,
55 Striped Sweet Pippin,
56 Summer Hagloe,
57 Smoke House
58 Sweet Romanite
59 Summer Queen.
60 { Sweet Red } June (on one plate.)
61 Swaar,
62 Talman's Sweet,
63 Twenty Ounce,
64 Vandever Pippin,
65 Wagener,
66 Walbridge
67 White Winter Pearmain,
68 " " Pippin,
69 Wine-Sap,
70 Willow Twig,
71 Yellow Bellflower.

Siberian Crab Apples.

1 Montreal Beauty,
2 { Small red, Black, } on one plate,
3 Transparent.
4 Hislop.

Pears.

1 Andrews,
2 Bartlett,
3 Belle Lucrative,
4 Beurre Brown,
5 " Clairgeau,
6 " d'Anjou,
7 " d'Aremberg,
8 " de Waterloo,
9 " Diel,
10 " Giffard,
11 " Hardy,
12 Bloodgood,
13 Brandywine
14 Buffum,
15 Doyenne Boussock,
16 " d'Alencon,
17 " d'Ete,
18 Duchesse d'Angouleme,
19 " d'Orleans,
20 Easter Beurre,
21 Flemish Beauty,
22 Glout Morceau,
23 Gray Doyenne,
24 Howell,
25 Lawrence,
26 Louise Bonne de Jersey,
27 Madeline,
28 Mt. Vernon,
29 Napoleon,
30 Nouveau Poiteau,
31 Osbands Summer,
32 Oswego Beurre,
33 Paradise d'Automne.
34 Pratt,
35 Rostiezer,
36 Seckel,
37 Sheldon,
38 Sterling,
39 Stevens Genesee,
40 Swans Orange,
41 Tyson,
42 Van Mons Leon le Clerk
43 Vicar of Winkfield,
44 White Doyenne,

Plums.

1 Bradshaw,
2 Coe's Golden Drop,
3 Dennison's Superb,
4 Duane's Purple,
5 Green Gage (Reine Claude,)
6 Jefferson,
7 Lawrence Favorite,
8 Lombard.
9 McLaughlin,
10 Peach,
11 Pond's Seedling,
12 Prune d'Agen,
13 Washington.

Cherries.

1 Black Tartarian,
2 Buttner's Yellow,
3 { Black Eagle, Early Richmond, May Duke } on one plate,
4 { Downer's Late Red, Monstreuse de Metzel, Napoleon Bigarreau, } on one plate,
5 Knight's Early Black,
6 Reine Hortense,
7 Sparhawk's Honey,
8 Tradescant's Black Heart,
9 White French Guigne,
10 Yellow Spanish,

Peaches.

1 Alberge Yellow,
2 Crawford's Early,
3 Crawford's Late,
4 Large Rareripe,
5 Peach Apricot,
6 Red-Cheek Melocoton,
7 Snow Peach,
8 Stump the World.

Strawberries.

1 Agriculturist
2 Austin Shaker.
3 Bicton Pine,
4 Genesee Seedling.
5 Hooker,
6 Hovey's Seedling,
7 McAvoy's Superior,
8 Monroe Scarlet.
9 Triomphe de Gand,
10 Trollop's Victoria,
11 Wilson's Albany Seedling,

Grapes.

1 Adirondac
2 Allens Hybrid
3 Catawba,
4 Chasselas Blanc,
5 Concord,
6 Clinton
7 Delaware,
8 Diana,
9 Eumelan.
10 Hartfort Prolific
11 Isabella,
12 Israella
13 Iona
14 Martha.
15 Rebecca.
16 Wilder, R. 4.
17 Salem,

Currants.

1 { Black Naples, Prince Albert, White Dutch, } on one plate,
2 Cherry,
3 White Grape,

Miscellaneos.

1 Com. European Walnut,
2 { Coryllus Avellana, Hazel Nuts, (six specimens,) } on one plate,
3 Dawnings Ever-bearing Mulberry
4 New Rochelle Blackberry,
5 Quince,

Raspberries.

1 { Red Antwerp, Yellow or White, } on one plate,
2 Doolittle Black Cap,
3 Mammoth Claster,

Gooseberries.

1 { Crown Rob, Whitesmith, } on one plate,
2 Houghton's Seedling.

Flowers & Shrubs.

1 African Tamarix,
2 Bignonia Radicans,
3 Chrysanthemums,
4 Crocus, (three specimens,)
5 Dahlia,
6 Daphne Mezereum,
7 Deuzia Scabra,
8 Dielytra Spectabilis,
9 { Double Crimson-Fl. Peach, " White " Almond,
10 { Double-fl'rg sweet Violets, Snow Drops,
11 Gladiolus
12 Hyacinths,
13 Hemerocalis Flava
14 Iris
15 Lillium lancifolium Rubrum,
16 Lonicera (3 specimens,)
17 Lonicera tartarica,
18 Magnolia (Umbrella Tree,)
19 Pyrus japonica
20 Snow Ball,
21 { Spirea Prunifolia, " Lanceolata,
22 " opulifolia,
23 " callosa,
24 { " Billardi, " Salicifolia,
25 Spirea Ulmaria Alba
26 " " Rosea
27 Tulips,
28 Viburnum lantana (Wayfaring Tree,)
29 Wiegelia Rosea.
30 White Fringe Tree

Roses.

1 Augusta,
2 " Mie,
3 Auretii,
4 Baronne Prevost,
5 Caroline de Sansel,
6 Geant des Batailles,
7 Hermosa
8 La Reine,
9 Leon des Combats,
10 Lord Raglan,
11 Madame Plantier,
12 Moss Rose,
13 Panache d'Orleans,
14 Queen of the Prairies,
15 Souvenir de la Malmaison,
16 Yellow Persian.

PRICE OF PLATES: Colored Fruits and Flowers 25 cts. each. | Colored Roses and Grapes 30 cts. each.

Catalogue of Fruit-and Flower-Plates. Drawn from Nature. Lithographed and Colored by Amana Society, Homestead P.O., Iowa, ca. 1870. Courtesy, The Albert R. Mann Library, New York State College of Agriculture and Life Sciences at Cornell University, Ithaca, N.Y.

This *Catalogue* is the second printed list of nurserymen's plates by Joseph Prestele and his son Gottlieb that has been located. It followed the *List* (ca. 1860) and, presumably, was issued after Joseph's death in 1867 but includes many plates that Joseph had created. That *List* offered 233 different plates; this one 237, which resulted from some new subjects having been added and a few old ones dropped. Prices were somewhat reduced, and for the first time it was stated that purchasers could have bound the assortment of plates they ordered.

Catalogue

OF

FRUIT- AND FLOWER- PLATES.

DRAWN FROM NATURE.

LITHOGRAPHED AND COLORED BY AMANA SOCIETY, HOMESTEAD P. O., IOWA.

Apples.

1 Alexander,
2 Astrachan Red,
3 Bailey's Sweet.
4 Baldwin,
5 Belmont,
6 Ben Davis (New York Pippin.)
7 Benoni,
8 Baltimore,
9 Black Apple
10 Canada Red,
11 Cooper,
12 Des Moines,
13 Domine,
14 Duchesse of Oldenburg.
15 Dutch Mignonne,
16 Eearly Harvest,
17 Early Joe,
18 Early Strawberry,
19 Early Pennock,
20 Esopus Spitzenburg,
21 Fameuse,
22 Fall Wine
23 Fulton
24 Gravenstein,
25 Grimes Golden
26 { Hubbardston Nonsuch, Michael Henry Pippin, } on one plate,
27 Jersey Sweet,
28 Jonathan,
29 Keswick Codlin,
30 King of Tompkins Co.,
31 Lady Apple,
32 Maidens Blush,
33 Minister,
34 Monmouth Pippin,
35 Mother,
36 Newtown Pippin,
37 Newtown Spitzenburg,
38 Northern Spy,
39 { Pecks Pleasant, White Bellflower, } on one plate,
40 Perry's Russet.
41 Pewaukee
42 Pomme Grise,
43 Porter,
44 Rambo,
45 Raule's Jannet,
46 Red Detroit,
47 Reinette Canada,
48 Rhode Island Greening,
49 Ribston Pippin,
50 Rome Beauty,
51 Roman Stem,
52 Roxbury Russet,
53 Seek-no-further,
54 St. Lawrence,
55 Striped Sweet Pippin,
56 Summer Hagloe,
57 Smoke House
58 Sweet Romanite
59 Sweet Bough,
60 Smith's Cider,
61 Summer Queen.
62 Sweet } June (on one plate.)
63 Red
64 Swaar,
65 Talman's Sweet,
66 Twenty Ounce,
67 Vandever Pippin,
68 Wagener,
69 Walbridge
70 White Winter Pearmain,
71 Wine-Sap,
72 Willow Twig,
73 Yellow Bellflower.

Siberian Crab Apples.

1 Montreal Beauty,
2 Small red,
3 Transcendant.
4 Hislop.
5 Telford.

Pears.

1 Bartlett,
2 Belle Lucrative,
3 Beurre Brown,
4 " Clairgeau,
5 " d'Anjou,
6 " d'Aremberg,
7 " de Waterloo,
8 " Giffard,
9 " Hardy,
10 Bloodgood,
11 Brandywine
12 Buffum,
13 Doyenne Boussock,
14 " d'Alencon,
15 " d'Ete,
16 Duchesse d'Angouleme,
17 " d'Orleans,
18 Easter Beurre,
19 Flemish Beauty,
20 Glout Morceau,
21 Gray Doyenne,
22 Howell,
23 Lawrence,
24 Louise Bonne de Jersey,
25 Madeline,
26 Mt. Vernon,
27 Napoleon,
28 Nouveau Poiteau,
29 Osbands Summer,
30 Oswego Beurre,
31 Paradise d'Automne.
32 Rostiezer,
33 Seckel,
34 Sheldon,
35 Sterling,
36 Stevens Genesee,
37 Swans Orange,
38 Tyson,
39 Van Mons Leon le Clerk
40 Vicar of Winkfield,
41 White Doyenne,
42 Winter Nelis.

Plums.

1 Bradshaw,
2 Coe's Golden Drop,
3 Dennison's Superb,
4 Duane's Purple,
5 Green Gage (Reine Claude,)
6 Jefferson,
7 Lawrence Favorite,
8 Lombard,
9 McLaughlin,
10 Peach,
11 Pond's Seedling,
12 Prune d'Agen,
13 Washington,
14 Miner.

Cherries.

1 Black Tartarian,
2 Buttner's Yellow,
3 Early Richmond,
4 { Downer's Late Red, Monstreuse de Metzel, Napoleon Bigarreau, } on one plate,
5 Knight's Early Black,
6 Reine Hortense,
7 Sparhawk's Honey,
8 Tradescant's Black Heart,
9 White French Guigne,
10 Yellow Spanish.
11 Governor's Wood.

Peaches.

1 Alberge Yellow.
2 Crawford's Early.
3 Crawford's Late,
4 Large Rareripe,
5 Peach Apricot,
6 Red-Cheek Melocoton,
7 Snow Peach,
8 Stump the World.
9 Haies Early.

Strawberries.

1 Agriculturist
2 Austin Shaker.
3 Bicton Pine,
4 Genesee Seedling,
5 Hooker,
6 Hovey's Seedling,
7 McAvoy's Superior,
8 Moore Scarlet.
9 Triomphe de Gand.
10 Trollop's Victoria.
11 Wilson's Albany Seedling,

Grapes.

1 Adirondac
2 Allens Hybrid
3 Catawba,
4 Chasselas Blanc,
5 Concord,
6 Clinton
7 Delaware,
8 Diana,
9 Eumelan.
10 Hartfort Prolific
11 Isabella,
12 Israella
13 Iona
14 Martha.
15 Rebecca.
16 Wilder, R. 4
17 Salem,

Currants.

1 Black Naples,
2 Cherry,
3 White Grape,

Miscellaneos.

1 Com. European Walnut,
2 { Coryllus Avellana, Hazel Nuts, (six specimens,) } on one plate,
3 Dawnings Ever-bearing Mulberry
4 New Rochelle Blackberry,
5 Quince,
6 Kittatinny,

Raspberries.

1 Red Antwerp,
2 Doolittle Black Cap,
3 Mammoth Cluster,
4 Yellow or White,

Gooseberries.

1 Crown Rob,
2 Houghton's Seedling.
3 White smith,

Flowers & Shrubs.

1 African Tamarix,
2 Bignonia Radicans,
3 Chrysanthemums,
4 Crocus, (three specimens,)
5 Dahlia,
6 Daphne Mezereum,
7 Deuzia Scabra,
8 Dielytra Spectabilis,
9 { Double Crimson-Fl. Peach, " White " Almond,
10 { Double-fl'rg sweet Violets, Snow Drops,
11 Gladiolus
12 Hyacinths,
13 Hemerocalis Flava
14 Iris
15 Lillium lancifolium Rubrum,
16 Lonicera (3 specimens,)
17 Lonicera tartarica,
18 Magnolia (Umbrella Tree,)
19 Pyrus japonica
20 Snow Ball,
21 { Spirea Prunifolia, " Lanceolata,
22 " opulifolia,
23 " callosa,
24 { " Billardi, " Salicifolia,
25 Spirea Ulmaria Alba
26 " " Rosea
27 Tulips,
28 Viburnum lantana (Wayfaring Tree,)
29 Wiegelia Rosea.
30 White Fringe Tree

Roses.

1 Augusta,
2 " Mie,
3 Auretii,
4 Baronne Prevost,
5 Caroline de Sansel,
6 Geant des Batailles,
7 Hermosa
8 La Reine,
9 Leon des Combats,
10 Lord Raglan,
11 Madame Plantier,
12 Moss Rose,
13 Panache d'Orleans,
14 Queen of the Prairies,
15 Souvenir de la Malmaison,
16 Yellow Persian.

PRICE OF PLATES: Colored Fruits and Flowers 25 cts. each, by 50 and more 15 cts. each.

The Binding of a Book of 50 Plates $ 1,00, a Book of 100 and more $ 2,00.

Catalogue of Fruit-And Flower-Plates. Drawn from Nature. Lithographed and Colored by Amana Society, Homestead, P.O., Iowa, ca. 1875. Courtesy, Mrs. Adolph Schmieder, Middle Amana, Iowa.

This is the third listing of nurserymen's plates by Joseph Prestele and his son Gottlieb that has been located. It has 244 different plates.

Fruits

Apples

ALEXANDER
(1) "*The Alexander Apple.*/By J. Prestele. Amana, Iowa Co., Iowa." (Watercolor; caption and signature written in ink.) Located: RU-2.
(2) "*The Alexander Apple.*/Lith. & col[d.] by Amana Society. Amana, Iowa County, Iowa." (Simplified version of #1.) Located: CU-1. Listed: CU-2.
(3) "*The Alexander Apple.*" (Revision of #2.) Located: CUM-1. Listed: CUM-2.
(4) Located: none. Listed: CvR-4.

AMERICAN PIPPIN
Located: none. Listed: CvR-4.

ASTRACHAN RED See RED ASTRACHAN

AUTUMN STRAWBERRY
(1) "*The Autumn Strawberry Apple.*/By J. Prestele, Iowa Co., Iowa." (Watercolor; caption and signature written in ink.) Located: RU-2.
(2) "*The Autumn Strawberry Apple.*" (Simplified version of #1.) Located: CU-1, CUM-1.
(3) Located: none. Listed: CvR-4.

AUTUMN SWAAR
Located: none. Listed: CvR-4.

BAILEY SWEET
(1) "*The Bailey Sweet Apple.*" Located: CU-1.
(2) "*The Bailey Sweet Apple.*" (Same as #1?) Located: RU-1.
(3) "*The Bailey Sweet Apple.*" (Simplified version of #1.) Located: OSU, CUM-1.
(4) Located: none. Listed as "Bailey's Sweet": CvR-4.

BALDWIN
(1) "*The Baldwin Apple.*/Lith & col[d.] by Amana Society. Homestead, Iowa." (Uncolored.) Located: AMAH.
(2) "*The Baldwin Apple.*/Lith & col[d.] by Amana Society. Amana, Iowa County, Iowa." (Same design as #1?) Located: CU-1.
(3) (Same design as #1?; colored.) Located: none. Listed: CUM-2.
(4) Located: none. Listed: CvR-4.

BALTIMORE
(1) "*Baltimore Apple.*" Located: AMAH.
(2) "*The Baltimore Apple.*" (Uncolored.) Located: AJM.

BALTIMORE RED
Located: none. Listed: CvR-4.

BEAUTY OF KENT
Located: none. Listed: CvR-4.

BELLFLOWER YELLOW See YELLOW BELLFLOWER, a synonym
Located: none. Listed: CvR-4.

BELMONT
(1) "*Belmont Apple.*/Amana Society by I. et G. Prestele. Amana, Iowa County, Iowa." Located: CU-1.
(2) (Same as #1?) Located: RU-2.
(3) Located: none. Listed: CUM-2.
(4) Located: none. Listed: CvR-4.

BEN DAVIS
(1) "*The Ben Davis Apple. (New York Pippin.)*" Located: OSU, CUM-1.
(2) Located: none. Listed: CvR-4.
(3) (Same as #1?) Located: none. Listed: CU-2.

BENONI
(1) "*The Benoni Apple.*" Located: OSU, CUM-1, CvR-2, RU-1, CU-1.
(2) Located: none. Listed: CvR-4.

BETHLEHEMITE
Located: none. Listed (as "Bethlemite"): CvR-4.

BLACK (see RED DETROIT, a synonym)
(1) "*The Black Apple.*/Lith. & col[d.] by Amana Society. Homestead, Iowa." (Printed from the same stone as "The Borovitsky Apple" plate: CU-1.) Located: CU-1.
(2) (Same as #1?) Located: none. Listed: CUM-2.

BLUE PEARMAIN
"*The Blue Pearmain Apple.*/By J. & G. Prestele, Amana, Iowa." (Watercolor; caption and signature written in ink.) Located: RU-2.

BOROVITSKY
(1) "*The Borovitsky Apple.*" (Printed from the same stone as the "Black Apple" plate; CU-1.) Located: RU-1, RU-2, CU-1. Listed (as "Borovitzky"): CU-2.
(2) "*Borovitsky.*" (Watercolor; caption written in ink. Simplified version of RU-2.) Located: APC.

BUCKINGHAM (See EQUINETELY and FALL QUEEN, synonyms)
Located: none. Listed: CvR-4.

CANADA RED (See RED CANADA)

CAROLINA RED JUNE (See RED JUNE, a synonym)
Located: none. Listed (as "Caroline Red June"): CvR-4.

CAYUGA REDSTREAK (See TWENTY OUNCE, a synonym)
Located: none. Listed: CvR-4.

CHENANGO STRAWBERRY
Located: none. Listed: CvR-4.

CITRON
"*The Citron Apple./[Zimmerman's?] Seedling.*" (Watercolor; caption written in ink.) Located: APC.

COLE'S QUINCE
"*Coles Quince.*" Located: OSU.

COOPER
(1) "*The Cooper Apple.*" Located: CU-1, RU-2.
(2) Located: none. Listed: CvR-4, CUM-2.

COURTHOUSE (See GILPIN, a synonym)
Located: none. Listed (as "Courthouse, or Little Red Romanite"): CvR-4.

CRANBERRY
"*The Cranberry Apple.*" Located: OSU.

DETROIT RED (See RED DETROIT, a synonym)
Located: none. Listed: CvR-4.

DOMINE
(1) "*Domine Apple.*" Located: OSU.
(2) "*Dominie.*" (Design different from #1.).) Located: CUM-1. Listed (as "Domine"): CUM-2.
(3) Located: none. Listed as "Domine": CvR-4.

DUCHESS OF OLDENBURGH
(1) "*Dutchesse of Oldenberg Apple.*" (Watercolor; caption written in ink.) Located: APC.
(2) "*The Dutchess of Oldenburg Apple.*" (Same design as #1.) Located: CU-1, CvR-2, RU-1. Listed (as "Duchesse of Oldenburg"): CU-2.
(3) "*The Dutchess of Oldenburg Apple.*" (A reissue, with minor changes, of #2?) Located: CUM-1. Listed (as "Duchesse of Oldenburg"): CUM-2.
(4) Located: none. Listed (as "Dutchess of Oldenburg"): CvR-4.

DUTCH MIGNONNE
(1) "*The Dutch Mignonne Apple.*" Located: CU-1.
(2) Located: none. Listed: CUM-2.

DYER
Located: none. Listed: CvR-4.

EARLY CHANDLER
Located: none. Listed: CvR-4.

EARLY GOLDEN SWEET
Located: none. Listed: CvR-4.

EARLY HARVEST
(1) "*The Early Harvest Apple.*" Located: CU-1, CUM-1.
(2) Located: none. Listed: CvR-4.

EARLY JOE
(1) "*The Early Joe Apple.*" Located: RU-2.
(2) "*The Early Joe Apple.*/Lith. & col^d.^ by Amana Society. Amana, Iowa County, Iowa." Located: CUM-1.
(3) (Same as #2?) Located: none. Listed: CU-2.
(4) Located: none. Listed: CvR-4.

EARLY MAY
Located: none. Listed: CvR-4.

EARLY PENNOCK
Located: none. Listed: CvR-4.

EARLY POUND (Not in Ragan, *Nomenclature of the Apple.*)
Located: none. Listed: CvR-4.

EARLY STRAWBERRY
(1) "*The Early Strawberry Apple.*" Located: APC.
(2) Located: none. Listed: CU-2.
(3) (Same as #2?) Located: none. Listed: CUM-2.
(4) Located: none. Listed: CvR-4. (Also incorrectly classified as a crab apple in CvR-4.)

EARLY SWEET BOUGH (A synonym of BOUGH; see SWEET BOUGH.)
Located: none. Listed: CvR-4.

EARLY TART
Located: none. Listed: CvR-4.

EARLY TRENTON (Not in Ragan, *Nomenclature of the Apple.*)
Located: none. Listed: CvR-4.

ENGLISH GOLDEN RUSSET (See GOLDEN RUSSET, a synonym)
Located: none. Listed: CvR-4.

EQUINETELY (See BUCKINGHAM, a synonym)
Located: none. Listed: CvR-4.

ESOPUS SPITZENBURG
(1) "*Esopus Spitzenburg Apple.*/Lith. & col^d.^ by Amana Society. Amana, Iowa County, Iowa." Located: CU-1, CUM-1. (D. M. Dewey published a theorem and freehand copy of this plate c.1859: CvR-3.)
(2) "*Esopus Spitzenburg Apple.*" Located: RU-2.
(3) Located: none. Listed: CvR-4.

FALLAWATER (See TULPEHOCKEN, a synonym)
Located: none. Listed: CvR-4.

FALL ORANGE
Located: none. Listed: CvR-4.

FALL PIPPIN
Located: none. Listed: CvR-4.

FALL QUEEN (See EQUINETELY and BUCKINGHAM, synonyms)
Located: none. Listed: CvR-4.

FALL STRIPE
Located: none. Listed: CvR-4.

FALL WINE
(1) "*The Fall Wine Apple.*" Located: AJM.
(2) "*The Fall Wine Apple.*/Lith. & col^d.^ by Amana Society. Amana, Iowa County, Iowa." Located: OSU. (Printed from the same stone, and with the same lettering, as #1, but with the addition of the Amana imprint.)
(3) "*Fall Wine.*/Lith. & col^d.^ by Amana Society, Amana, Iowa County, Iowa." (Printed from the same stone as #2, but with differently lettered caption.) Located: CU-1.
(4) (Same impression as #3?) Located: none. Listed: CvR-4.
(5) "*Fall Wine.*/Lith. & col^d.^ by Amana Society. Amana, Iowa County, Iowa." (Same impression as #3?) Located: RU-2. (Printed from the same stone as the "Golden Reinette" plate; CU-1.)
(6) Located: none. Listed (as "Fall Wine"): CvR-4.

FALL WINESAP
Located: none. Listed: CvR-4.

FAMEUSE (See SNOW, a synonym)
(1) "*The Fameuse Apple.*" Located: CU-1, OSU.
(2) Located: none. Listed: CvR-4.

FOURTH OF JULY
Located: none. Listed: CvR-4.

FULTON
(1) "*The Fulton Apple.*" Located: CU-1, OSU,

CUM-1.
(2) "*The Fulton Apple.*" (Same as #1?) Located: RU-2.
(3) Located: none. Listed: CvR-4.

GENESEE CHIEF
(1) "*The Genesee Chief Apple.*" (Watercolor; caption written in ink.) Located: RU-2.
(2)"*The Genesee Chief Apple.*" Located: CU-1.
(3) (Same design as #2?) Located: none. Listed: CUM-2.

GEORGIA JUNE (See RED JUNE, a synonym)
Located: none. Listed: CvR-4.

GILPIN (See COURTHOUSE, a synonym)
Located: none. Listed: CvR-4.

GOLDEN REINETTE
(1) "*Golden Reinette.*" Located: CU-1.
(2) "*Golden Reinette.*/Lith. & col[d.] by Amana Society. Amana, Iowa County, Iowa." Located: RU-2. (Printed from the same stone as the "Fall Wine" plate; RU-2.)

GOLDEN RUSSET (See ENGLISH GOLDEN RUSSET, a synonym)
(1) "*Golden Russet Apple.*" Located: CU-1.
(2) (Same as #1?) Located: none. Listed: CUM-2.
(3) Located: none. Listed: CvR-4.

GOLDEN SEEDLING
Located: none. Listed: CvR-4.

GRAVENSTEIN
(1) "*The Gravenstein Apple.*/Lith. & col[d.] by G. Prestele. Ebenezer, near Buffalo, N.Y." Located: RU-1, CvR-2.
(2) "*Gravenstein Apple.*/Sold Only by D. M. Dewey, Rochester, N.Y." [Printed by E. B. & E. C. Kellogg, Hartford, Conn.?] Located: CvR-3.
(3) "*The Gravenstein Apple.*/Lith. & col[d.] by Amana Society. Amana, Iowa County, Iowa." (Printed from the same stone, with the same lettered caption as #1, but with the addition of the Amana imprint.) Located: CU-1.
(4) Same as #3? Located: none. Listed: CUM-2.
(5) Located: none. Listed: CvR-4.

GRIMES GOLDEN
(1) "*The Grimes' Golden Pippin Apple.*" Located: CUM-1.
(2) Located: none. Listed (as "Grimes' Golden Pippin"): CvR-4.

HAAS (See FALL QUEEN, a synonym)
Located: none. Listed (as "Haas or Gros Pommier") CvR-4.

HASKELL SWEET
Located: none. Listed: CvR-4.

HERR'S JUNE
Located: none. Listed: CvR-4.

HIGHLAND BEAUTY
"*Highland Beauty*/Drawn from Nature Lith. & Col[d.] by W. H. Prestele, Iowa City, Iowa." (Produced after 1879? Not in Prestele's 1879 catalogue.) Located: AES.

HOMINY (See SOPS OF WINE, a synonym)
Located: none. Listed: CvR-4.

HORSE (See HAAS, a synonym)
Located: none. Listed: CvR-4.

HOWELL
"*The Howell Apple.*" Located: AJM. Listed: none. (Plate published after the last J. and/or G. catalogues were issued, ca. 1875?)

HUBBARDSTON NONSUCH. MICHAEL HENRY PIPPIN
(1) "I *The Hubbardston's Nonsuch.* II *Michael Henry Pippin Apple.*" Located: RU-2, CU-1.
(2) (Same design as #1?) Located: (uncolored), AMAH.
(3) Located: none. Listed: CvR-4.

HUNTSMAN'S FAVORITE
Located: none. Listed: CvR-4.

JERSEY SWEET
(1) "*The Jersey Sweet Apple.*" Located: CU-1.
(2) (Same as #1?) Located: none. Listed: CUM-2.

JONATHAN
(1) "*Jonathan Apple.*/Lith. & col[d.] by Amana Society. Amana, Iowa County, Iowa." Located: RU-2, CU-1, CUM-1.
(2) Located: none. Listed: CvR-4.

KENTUCKY CREAM
Located: none. Listed: CvR-4.

KENTUCKY QUEEN (See BUCKINGHAM, a synonym)
Located: none. Listed: CvR-4.

KENTUCKY REDSTREAK (See BEN DAVIS, a synonym)
Located: none. Listed: CvR-4.

KESWICK CODLIN
(1) "*Keswick Codlin Apple.*/Lith. & col[d.] by Amana Society. Amana, Iowa County, Iowa." Located: CU-1. Listed: CUM-2, CU-2. (A theorem and freehand version of this design published by D. M. Dewey, of c.1859, with his printed caption and imprint. Located: CvR-3.)
(2) "*The Keswick Codlin Apple.*" Located: AJM.
(3) Located: none. Listed: CvR-4.

KING APPLE
"*The King Apple.*" Located: RU-1.

KING OF PIPPINS
(1) "*The King of Pippin Apple.*/By J. Prestele. Amana, Iowa Co., Iowa." (Watercolor; caption written in ink.) Located: RU-2.
(2) "*The King of Pippin Apple.*" (Revised and simplified version of #1.) Located: CU-1. Listed: CU-2.

KING OF TOMPKINS COUNTY
(1) Located: none. Listed: CU-2.
(2) Located: none. Listed: CUM-2.
(3) Located: none. Listed: CvR-4.

KIRKBRIDGE WHITE
Located: none. Listed: CvR-4.

LADY
(1) "*The Lady Apple.*/Lith. & col^d. by Amana Society. Amana, Iowa County, Iowa." Located: CU-1.
(2) (Same design as #1?) Located: none. Listed: CUM-2.
(3) Located: none. Listed: CvR-4.

LADY FINGER
Located: none. Listed: CvR-4.

LARGE STRIPED PEARMAIN
Located: none. Listed: CvR-4.

LAWVER
Located: none. Listed: CvR-4.

LIMBERTWIG
Located: none. Listed: CvR-4.

LOWELL
Located: none. Listed: CvR-4.

MAIDEN BLUSH
(1) "*The Maiden's Blush Apple.*" Located: CU-1, CUM-1.
(2) Located: none. Listed: CvR-4. (Also classified as a crab apple.)

MANAGERE (Not in Ragan, *Nomenclature of the Apple.*)
"*The Managere Apple.*/Drawn from Nature by J. & G. Prestele. Amana Society." (Watercolor; caption and signature written in ink.) Located: RU-2.

MELON
(1) "*The Melon Apple.*" Located: RU-2.
(2) "*Melon Apple.*" Located: CU-1. (This is a slightly altered version of the Joseph Prestele "Melon Apple" plate in the *Horticulturist*, n.s., vol. 4 [1854].)
(3) Located: none. Listed: CvR-4.

MINISTER
(1) "*Minister apple.*" (Watercolor; caption written in ink.) Located: APC.
(2) "*The Minister Apple.*" Located: RU-1, RU-2.
(3) "*The Minister Apple.*" Located: CU-1, CvR-2.
(4) Located: none. Listed: CUM-2.

MISSOURI PIPPIN
Located: none. Listed: CvR-4.

MISSOURI SUPERIOR
Located: none. Listed: CvR-4.

MONMOUTH PIPPIN
(1) "*The Monmouth Pippin.*" Located: CU-1.
(2) (Same design as #1?) Located: none. Listed: CUM-2.

MOTHER
(1) "*The Mother Apple.*/Lith. & col^d. by Amana Society. Amana, Iowa County, Iowa." Located: CUM-1, CU-1.
(2) Located: none. Listed: CvR-4.

NEWTOWN PIPPIN
(1) "*The Newtown Pippin.*" Located: RU-2, CU-1.
(2) Located: none. Listed: CUM-2.

NEWTOWN SPITZENBURG
(1) "*Newtown Spitzenburg Apple.*" Located: RU-2, CU-1.
(2) Located: none. Listed: CUM-2.

NEW YORK PIPPIN
Located: none. Listed: CvR-4.

NORTHERN SPY
(1) "*Northern Spy Apple.*/J. Prestele, Lith. & Painter. Ebenezer, n. Buffalo, N.Y." Located: RU-2, CvR-2. D. M. Dewey issued this same plate c.1859 without the Prestele imprint and with his printed caption and imprint (removed by trimming from the CvR-3 example).
(2) "*Northern Spy Apple.*" (A varient of #1.) Located: CU-1.
(3) "*The Northern Spy Apple.*" (A variant of #2.) Located: CUM-1.
(4) Located: none. Listed: CvR-4.

OHIO, OR MYER'S NONPARELL
Located: none. Listed: CvR-4.

ORTLEY, or WHITE BELLFLOWER (See WHITE BELLFLOWER.)
Located: none. Listed: CvR-4.

PECK'S PLEASANT. WHITE BELLFLOWER
(1) "I *The Peck's Pleasant.* II *White Bellflower Apple.*" Located: CU-1, CUM-1.
(2) Located: none. Listed: CvR-4.

PENNSYLVANIA CIDER
Located: none. Listed: CvR-4.

PENNSYLVANIA REDSTREAK
Located: none. Listed: CvR-4.

PERRY RUSSET (See GOLDEN RUSSET, a synonym)
(1) "*The Perry Russet Apple.*" Located: CUM-1.
(2) Located: none. Listed: CvR-4.

PEWAUKEE
(1) "*The Pewaukee Apple.*/Grown by Geo. P. Peffer, Pewaukee, Wis./Lith. & col^d. by Amana Society, Homestead, Iowa." Located: CUM-1.
(2) Located: none. Listed (as written addition to Prestele's 1879 catalogue): CvR-4.

PLUMB'S CIDER
Located: none. Listed: CvR-4.

POMME GRIS
(1) "*The Pomme Grise Apple.*" Located: CU-1.
(2) (Same as #1?) Located: none. Listed: CUM-2.
(3) Located: none. Listed (as "Pomme Griss"): CvR-4.

PORTER
(1) "*The Porter Apple.*/By J. Prestele. Amana Society, Iowa." (Watercolor; caption and signature written in ink.) Located: RU-1.
(2) "*The Porter Apple.*/Lith. & col^d. by Amana Society. Amana, Iowa County, Iowa." (Same design as #1?) Located: CU-1.

(3) (Same design as #1?) Located: none. Listed: CUM-2.
(4) Located: none. Listed: CvR-4.

PRIMATE
Located: none. Listed: CvR-4.

PRYOR'S RED
Located: none. Listed: CvR-4.

PUMPKIN RUSSET
Located: none. Listed: CvR-4.

QUEEN OF THE WEST
Located: none. Listed: CvR-4.

RAMBO (See SEEK-NO-FURTHER, a synonym)
(1) "*The Rambo Apple.*/Lith. & col^d. by Amana Society. Amana, Iowa County, Iowa." Located: CU-1.
(2) "*Rambo Apple.*"/D. M. Dewey's Series of Fruits, Flowers, and Ornamental Trees, Rochester, N.Y./Lith. of E. B. & E. C. Kellogg. Hartford, Conn." (Same design as #1, but reversed.) Located: CvR-3.
(3) (Same as #1?) Located: none. Listed: CUM-2.
(4) Located: none. Listed: CvR-4.

RAMSDELL SWEETING
Located: none. Listed: CvR-4.

RALLS JANET
(1) "*The Rawles Janet Apple.*" Located: CU-1. Listed (as "Raule's Janet"): CU-2.
(2) "*The Rawle's Jannet Apple.*" (Design different from #1.) Located: OSU, CUM-1. Listed (as "Raule's Jannet"): CUM-2.
(3) Located: none. Listed (as "Rowles Janet"): CvR-4.

RED ASTRACHAN (See ASTRACHAN RED, a synonym.)
(1) "*The Red Astrachan Apple.*" This is a variant, perhaps later in date, of Joseph Prestele's plate in the *Horticulturist*, vol. 5, n.s. (October 1855). Located: OSU, CUM-1.
(2) "*The Red Astrachan Apple.*" (Same design as #1?; uncolored.) Located: AS.
(3) "*Red Astrachan.*" Located: RU-1.
(4) "*Red Astrachan.*" (Variant of #1.) Located: CU-1.
(5) Located: none. Listed (as "Red Astrachan"): CvR-4.

RED CANADA (See CANADA RED, a synonym.)
(1) "*The Red Canada Apple.*" Located: CU-1.
(2) Located: none. Listed: CUM-2.
(3) Located: none. Listed (as "Canada Red"): CvR-4.

RED DETROIT (See DETROIT RED and BLACK, synonyms.)
(1) "*Red Detroit Apple.*" Located: RU-2.
(2) "*Red Detroit Apple.*" (Same design as #1?) Located: CU-1, CUM-1.

RED JUNE—(See CAROLINA RED JUNE, synonym.)
Located: none. Listed: CvR-4.

REINETTE CANADA (See CANADA REINETTE, a synonym.)
(1) "*The Reinette Canada Apple.*" Located: CU-1.
(2) (Same as #1?) Located: none. Listed: CUM-2, AAS-3.

RHODE ISLAND GREENING
(1) "*The Rhode Island Greening.*/Lith. & col^d. by Amana Society. Amana, Iowa County, Iowa." Located: CU-1.
(2) Located: none. Listed: CUM-2.
(3) "*Rhode Island Greening Apple.*" (Caption written in ink. Variant of #1.) Located: APC.
(4) "*Rhode Island Greening Apple.*/D. M. Dewey's Series of Fruits, Flowers, and Ornamental Trees, Rochester, New York." (Printed by E. B. & E. C. Kellogg, Hartford, Conn.? Earlier than #1 whose design is a variant.) Located: CvR-3.
(5) Located: none. Listed: CvR-4.

RIBSTON PIPPIN
(1) "*The Ribston Pippin.*" (Watercolor; caption written in ink.) Located: RU-2.
(2) "*The Ribston Pippin Apple.*/Lith. & Col^d. by Amana Society. Amana, Iowa County, Iowa." (Variant of #1; minor deletions.) Located: CU-1.
(3) (Same as #2?) Located: none. Listed: CUM-2.

ROMAN STEM
(1)"*The Roman Stem Apple.*" Located: RU-2.
(2) "*The Roman Stem Apple.*" (Same as #1?) Located: CU-1, CUM-1.
(3) Located: none. Listed: CvR-4.

ROME BEAUTY
(1) "*Rome Beauty Apple.*" Located: CU-1.
(2) (Same as #1?) Located: none. Listed: CUM-2.
(3) Located: none. Listed: CvR-4.

ROXBURY RUSSET
(1) "*The Roxbury Russet.*" Located: CU-1.
(2) (Same as #1?) Located: none. Listed: CUM-2.
(3) Located: none. Listed: CvR-4.

ST. LAWRENCE
(1) "*The St. Lawrence Apple.*" Located: RU-1.
(2) "*The St. Lawrence Apple.*/Lith. & Col^d. by Amana Society. Amana, Iowa County, Iowa." Located: CU-1, CUM-1.
(3) Located: none. Listed: CvR-4.

SAXTON (See FALL STRIPE, a synonym.)
Located: none. Listed: CvR-4.

SEEK-NO-FURTHER (See RAMBO, a synonym.)
(1) "*Seek no Further.*" Located: RU-2.
(2) "*Seek no Further.*" Located: CU-1. (This design copied in theorem and freehand by D. M. Dewey for a plate in his 1859 collection: CvR-3.)
(3) Located: none. Listed: CvR-4.

SHOCKLEY
Located: none. Listed: CvR-4.

SMITH'S CIDER
Located: none. Listed: CvR-4.

SMOKEHOUSE
(1) "*The Smokehouse Apple.*/Lith. by Amana So-

ciety, Homestead, Iowa." Located: RU-2, CU-1.
(2) (Same as above?) Located: none. Listed: CUM-2.
(3) Located: none. Listed: CvR-4.

SNOW (See FAMEUSE, a synonym)
Located: none. Listed: CvR-4.

SOPS-OF-WINE (See HOMINY, a synonym)
Located: none. Listed: CvR-4.

STRIBLING
Located: none. Listed: CvR-4.

"STRIPED RED PIPPIN" (Not in Ragan, *Nomenclature of the Apple*)
Located: none. Listed (as above): CvR-4.

STRIPED SWEET PIPPIN
(1) "Striped Sweet Pippin." Located: CU-1.
(2) (Same as #1?) Located: none. Listed: CUM-2.

SUMMER HAGLOE
(1) "*Summer Hagloe Apple.*" (Watercolor; caption written in ink.) Located: APC.
(2) "*The Summer Hagloe Apple.*/J. Prestele." (Reproduced from #1 with minor variations.) Located: RU-2, CU-1, CUM-1, RU-1. (A plate, printed from the same stone as #2, and with the "J. Prestele" signature, was captioned "The Sweet Romanite Apple"; see entry.)

SUMMER KING
Located: none. Listed: CvR-4.

SUMMER PEARMAIN
Located: none. Listed: CvR-4.

SUMMER QUEEN
(1) "*The Summer Queen Apple.* Lith. & Col^d. by Amana Society, Homestead, Iowa." Located: AES.
(2) "*Summer Queen.*" Located: RU-2.
(3) "*Summer Queen.*" (Same as #2?) Located: none. Listed: CU-2, CUM-2.
(4) Located: none. Listed: CvR-4.

SUMMER ROSE
(1) "*Summer Rose Apple.*" (Watercolor; caption written in ink.) Located: RU-1.
(2) "*The Summer Rose Apple.*" (Watercolor? Caption handwritten.) Located: AES.
(3) "*Summer Rose Apple.*" Located: RU-1, RU-2.
(4) "*Summer Rose Apple.*" Located: CU-1.

SWAAR
(1) "*The Swaar Apple.*" Located: OSU. (D. M Dewey copied this design in a theorem and freehand plate for his 1859 collection: CvR-3.)
(2) "*The Swaar Apple.*" Located: RU-2.
(3) Located: none. Listed: CU-2, CUM-2.
(4) Located: none. Listed: CvR-4.

SWEET BOUGH (See EARLY SWEET BOUGH, also a synonym of BOUGH.)
"*The Sweet Bough Apple.*" Located: AJM.

SWEET JUNE
Located: none. Listed: CvR-4.

"SWEET JUNE. RED JUNE."
"1 *The Sweet June Apple.* 2 *The Red June.*" (On one plate.) Located: CU-1, CUM-1.

SWEET ROMANITE
(1) "*The Sweet Romanite Apple.*/J. Prestele." (This design was also used for "The Summer Hagloe Apple"; see entry.) Located: RU-2.
(2) "*The Sweet Romanite Apple.*/Lith. & Col^d. by Amana Society, Homestead, Iowa." (Printed from the same stone as #1, different lettering.) Located: CU-1.
(3) (Same as #2?) Located: none. Listed: CUM-2.

TETOFSKI (See TETOFSKY and RUSSIAN CRAB, synonyms.)
(1) "*The Tetofsky Apple.*" Located: private collection.
(2) Located: none. Listed (as "Tetofsky," a crab apple): CvR-4.

TOLMAN'S SWEET (TOLMAN, a synonym.)
(1) "*The Talman's Sweeting Apple.*" Located: CU-1, CUM-1. Listed (as "Talman's Sweet"): CU-2, CUM-2.
(2) Located: none. Listed (as "Talman's Sweet") CvR-4.

TULPEHOCKEN (See FALLAWATER, a synonym.)
Located: none. Listed: CvR-4.

TWENTY OUNCE (See CAYUGA REDSTREAK, a synonym.)
(1) "*The Twenty Ounce Apple.*" Located: CU-1.
(2) (Same as #1?) Located: AMAH.

UTTER'S LARGE RED
Located: none. Listed: CvR-4.

VANDERVERE PIPPIN (See SMOKEHOUSE, a synonym.)
(1) "*The Vandevere Pippin Apple.*" Located: RU-1.
(2) "*The Vandevere Pippin Apple.*" (Same as #1?) Located: CU-1.
(3) (Same as #1?) Located: none. Listed: CUM-2.
(4) Located: none. Listed: CvR-4.

VIRGINIA JUNE
Located: none. Listed: CvR-4.

WAGENER
(1) "*The Wagener Apple.*" Located: CU-1.
(2) "*The Wagener Apple.*" (Slight revision of #1, including redrawn caption.) Located: CUM-1.
(3) Located: none. Listed: CvR-4.

WALDBRIDGE
(1) "*The Waldbridge Apple.*" Located: CUM-1.
(2) Located: none. Listed: CvR-4.

WEALTHY
Located: none. Listed: CvR-4.

WEYAUWEGA
Located: none. Listed (as written addition to 1879 catalogue): CvR-4.

WHITE PIPPIN
Located: none. Listed: CvR-4.

"WHITE POUND" (Not in Ragan, *Nomenclature of the Apple*)
Located: none. Listed: CvR-4.

WHITE WINTER PEARMAIN (MICHAEL HENRY, a synonym.)
(1) "*White Winter Pearmain Apple.* Lith. & Col[d.] by Amana Society. Amana, Iowa County, Iowa." Located: CU-1, CUM-1.
(2) Located: none. Listed: CvR-4.

WHITE WINTER PIPPIN
(1) "*White Winter Pippin.*" Located: CU-1.
(2) (Same as #1?) Located: none. Listed: CUM-2.

WILLIAMS' FAVORITE
Located: none. Listed: CvR-4.

WILLOW TWIG
(1) "*Willow Twig Apple.*" Located: OSU.
(2) "*The Willow Twig Apple.*" (Different design from #1.) Located: CUM-1.
(3) Located: none. Listed: CvR-4.

WINESAP
(1) "*The Wine Sap Apple.*" (Watercolor; caption written in ink.) Located: APC.
(2) "*The Wine Sap Apple.*" (Slightly simplified version of #1.) Located: OSU, CU-1, CUM-1.
(3) "*The Wine Sap Apple.*" (Same design as #2?) Located, uncolored: AHSM.
(4) Located: none. Listed: CvR-4.

WINTER
Located: none. Listed (as "Winter, or Myer's May"): CvR-4.

WINTER SWEET PARADISE
Located: none. Listed: CvR-4.

WINTER WINE
Located: none. Listed: CvR-4.

WOLF RIVER
Located: none. Listed (as written addition to 1879 catalogue): CvR-4.

YELLOW BELLFLOWER (See BELLFLOWER YELLOW, a synonym.)
"*The Yellow Bellflower Apple.* Lith. & col[d.] by Amana Society. Amana, Iowa County, Iowa." Located: CU-1, CUM-1, OSU.

YELLOW TWIG
"*The Yellow Twig Apple.*" Located: AJM.

Apricots

BREDA
Located: none. Listed: CvR-4.

GOLDEN
Located: none. Listed: CvR-4.

MOORPARK
Located: none. Listed: CvR-4.

PEACH
Located: none. Listed: CvR-4.

Blackberries

KITTATINNY
(1) "*The Kittatinny Blackberry.*" Located: APC.
(2) "*The Kittatinny Blackberry.*" (Simplified version of #1; compactly designed, possibly intended for an illustration.) Located: APC.
(3) Located: none. Listed: CvR-4.

LAWTON
Located: none. Listed: CvR-4.

MISSOURI MAMMOTH
Located: none. Listed: CvR-4.

NEW ROCHELLE
(1) "*New Rochelle Blackberry.*/Lith. & col[d.] by Amana Society. Amana, Iowa County, Iowa." Located: CU-1.
(2) (Same as #1?) Located: none. Listed: CUM-2.

SNYDER
Located: none. Listed (as written addition to W. H. Prestele's 1879 catalogue): CvR-4.

WHITE
"*White Blackberry.*" Located: CU-1.

WILSON'S EARLY
(1) "*The Wilson's Early Blackberry.*/Introduced by R. O. Thompson, Nursery Hill, Nebraska./Lith. & col[d.] by Amana Society. Amana, Iowa County, Iowa." Located CU-1. Listed: none.
(2) Located: none. Listed: CvR-4.

Cherries

BELLE MAGNIFIQUE
Located: none. Listed: CvR-4.

BIGARREAU or YELLOW SPANISH (See YELLOW SPANISH, a synonym.)
Located: RU-1, CvR-2.

BLACK EAGLE. EARLY RICHMOND. MAY DUKE (On one plate) (See "MAYDUKE BLACK EAGLE EARLY RICHMOND CHERRY.")
(1) Located: none. Listed: CU-2, CUM-2.
(2) Located: none. Listed: CvR-4.

BLACK TARTARIAN
(1) "*Black Tartarian Cherry.*" Located: CU-1. (A theorem and freehand variant of this design was issued by "D. M. Dewey, Rochester, N.Y.," ca. 1859, with the printed caption "*Black Tartarian.*")
(2) (Same as #1?) Located: none. Listed: CUM-2.
(3) "*The Black Tartarian Cherry.*" (Variant of #1.) Located: APC.

BUTTNER'S YELLOW
(1) "*The Buttner's Yellow Cherry.*" Located: CvR-2.
(2) "*The Buttner's Yellow Cherry.*/Lith. & col[d.] by Amana Society. Amana, Iowa County, Iowa." (Variant of #1.) Located: CU-1.
(3) (Same as #1?) Located: none. Listed: CUM-2.
(4) "*The Buttner's Yellow Cherry.*" (Same as #2?) Located: RU-1, RU-1.

"CHERRIES./I NAPOLEON BIGARREAU. II MONSTREUSE DE MEZEL III DOWNER'S LATE RED."
Located: RU-1, CU-1, CUM-1.

"DOWNER'S LATE RED. MONSTREUSE DE METZEL [*sic*]. NAPOLEON BIGARREAU (On one plate.)"
(Same design as the preceding entry?) Located: none. Listed: CU-2, CUM-2.

"EARLY PURPLE GUIGNE"
Located: none. Listed: CvR-4.

"EARLY RICHMOND"
(1) Located: none. Listed: CvR-4.
(2) "Early Richmond." (Same as #1?) Listed: CvR-5.

"ELKHORN"
Located: none. Listed: CvR-4.

"ELTON"
Located: none. Listed: CvR-4.

"ENGLISH MORELLO"
Located: none. Listed: CvR-4.

GERMAN MORELLO
"*The german Morello./Cherry.*" (Watercolor; caption written in ink.) Located: APC.

GOVERNOR WOOD
(1) Located: none. Listed: CvR-4.
(2) Located: none. Listed (as "Governor's Wood"): AAS-3.

KENTISH
Located: none. Listed: CvR-4.

KNIGHT'S EARLY BLACK
(1) "*Knight's Early Black./Cherry.*" (Watercolor; caption written in ink.) Located: APC.
(2) "*Knight's Early Black.*" (Slightly simplified version of #1.) Located: CvR-2. (Slightly modified version in the *Horticulturist*, vol. 6, n.s. [October 1856].)
(3) "*The Knight's Early Black Cherry.*" (A revision, apparently of #2.) Located: CU-1.
(4) (Same as #3?) Located: none. Listed: CUM-2.
(5) (Same as #2?) Located: RU-1.
(6) Located: none. Listed: CvR-4.

LUELLING
Located: none. Listed: CvR-4.

MAYDUKE. BLACK EAGLE. EARLY RICHMOND. (See entry, BLACK EAGLE. . . .)
"I *Mayduke* II *Black Eagle* III *Early Richmond Cherry.*" Located: CU-1.

MONTGOMERY
Located: none. Listed: CvR-4.

NAPOLEON BIGARREAU
"*Napoleon Bigarreau./Cherry.*" (Watercolor; caption written in ink.) Located: APC.

OHIO BEAUTY
Located: none. Listed: CvR-4.

OLIVET
Located: none. Listed: CvR-4.

REINE HORTENSE
(1) "*The Reine Hortense Cherry.*" Located: CvR-2.
(2) "*The Reine Hortense Cherry.*" (Simplified version of #1; identical hand-lettered caption.) Located: CU-1.
(3) (Same as #2?) Located: none. Listed: CUM-2.
(4) (Same as #2?) Located: RU-1.
(5) Located: none. Listed: CvR-4.

SPARHAWK'S HONEY
(1) "*Sparhawk's Honey/Cherry.*" (Watercolor; caption written in ink.) Located: APC.
(2) "*The Sparhawks Honey Cherry.*" (Same design as #1.) Located: CvR-2.
(3) "*The Sparhawks Honey Cherry.* Lith. & col[d.] by Amana Society. Amana, Iowa County, Iowa." (Same design as #1.) Located: CU-1.
(4) (Same as #3?) Located: none. Listed: CUM-2.
(5) (Same as #2?) Located: RU-1, RU-2.
(6) Located: none. Listed: CvR-4.

TRADESCANT'S BLACK HEART
(1) "*The Tradescant's Black Heart Cherry.*" Located: CvR-2, CU-1.
(2) (Same as #1?) Located: RU-1.
(3) (Same as #1?) Located: none. Listed: CUM-4.
(4) Located: none. Listed: CvR-4.

UTAH HYBRID
Located: none. Listed: CvR-4.

WHITE FRENCH GUIGNE
(1) "*The White French Guigne Cherry.*" Located: CU-1, RU-1, RU-2.
(2) (Same as #1?) Located: none. Listed: CUM-2.

YELLOW SPANISH (See also BIGARREAU OR YELLOW SPANISH)
(1) "*The Yellow Spanish Cherry.*/Eng. & col[d.] by G. Prestele," Located: CU-1.
(2) (Same as #1?) Located: none. Listed: CUM-2.

YELLOW SPANISH, MONSTREUSE DE MEZEL. DOWNER'S LATE RED (On one plate)
Located: none. Listed: CvR-4.

Crab Apples

AUCUBIFOLIA
Located: none. Listed (as "Ancubifolia"): CvR-4.

BEECHER'S SWEET
Located: none. Listed (as written addition to W.H. Prestele's 1879 catalogue): CvR-4.

BRIER'S SWEET CRAB
Located: none. Listed (as "Briar's Sweet"): CvR-4.

CHICAGO
Located: none. Listed: CvR-4.

GENERAL GRANT
Located: none. Listed (as written addition to W. H. Prestele's 1879 catalogue): CvR-4.

GOLDEN BEAUTY
Located: none. Listed: CvR-4.

HYSLOP
(1) "*Hyslops Siberian Crab Apple*. Lith. & col. by Amana Society. Amana Iowa County Iowa." (Printed from the same stone as the "Transparent Siberian Crab Apple" plate, altered caption; see entry.) Located: OSU.
(2) "*The Hyslop Crab Apple*." (Completely different design from #1.) Located: CUM-1.
(3) Located: none. Listed: CvR-4.

KENYON
Located: none. Listed: CvR-4

LADY ELGIN
Located: none. Listed: CvR-4.

LARGE RED
"*The Large Red Crab Apple*." (The same design was used for printing "The Montreal Beauty Crab Apple" plate; see entry.) Located: OSU.

LARGE RED SIBERIAN
Located: none. Listed: CvR-4.

LATE WINTER SIBERIAN
Located: none. Listed (as "Lake Winter Siberian"): CvR-4.

LARGE YELLOW SIBERIAN
Located: none. Listed: CvR-4.

MARENGO
Located: none. Listed: CvR-4.

MEADERS WINTER
Located: none. Listed: CvR-4.

MINNESOTA
Located: none. Listed: CvR-4.

MONTREAL BEAUTY
(1) "*The Montreal Beauty Crab Apple*." (This plate was also issued with the caption "The Large Red Crab Apple"; see entry.) Located: CUM-1, CU-1.
(2) Located: none. Listed: CvR-4.

QUAKER BEAUTY
Located: none. Listed: CvR-4.

SIBERIAN CRAB APPLE
(1) "*The siberian Crab Apples./1. Small Red. 2. Transparent or Yellow. 3. Black*." (Watercolor; caption written in ink. A preliminary sketch, without foliage, for the plates below.) Located: APC.
(2) "*siberian Crab Apple*." (Watercolor? Handwritten caption. Same design as #1?) Located: AES.
(3) "*The Small Red Siberian Crab Apple*. Lith. & col[d.] by Amana Society, Homestead, Iowa." (Same design as the upper half of #3 without the background of outlines of leaves.) Located: AES.
(4) "I. *Small Red* II *Black/Siberian Crab Apple*." Located: RU-2, CU-1, CUM-1.

SNYDER
Located: none. Listed: CvR-4.

SOULARD
Located: none. Listed: CvR-4.

"STRIPED WINTER" (Not in Ragan, *Nomenclature of the Apple*.)
Located: none. Listed: CvR-4.

SYLVAN SWEET
Located: none. Listed: CvR-4.

TELFER (See TELFORD.)

TELFER'S ORANGE
Located: none. Listed: CvR-4.

TELFER'S SWEET
Located: none. Listed: CvR-4.

TELFORD [TELFER]. (See TELFER SWEET, a synonym.)
(1) "*The Telford Crab*./Painted by G. Prestele, Amana, Iowa." (Watercolor? Handwritten signature.) Located: AS.
(2) "*The Telford Crab*." Located: AEG.
(3) Located: none. Listed (as "Telford"): AAS-3.

TRANSCENDENT
Located: none. Listed: CvR-4.

TRANSPARENT
"*Transparent Siberian Crab Apple*." Lith. & col. by Amana Society Amana Iowa County Iowa." (Sometime after this plate was published, the caption on the stone was altered, deleting "Transparent" and substituting "Hyslops," the corrected plate then reading "Hyslops Siberian Crab Apple"; see entry.) Located: CUM-1.

WHITNEY'S NO. 20
Located: none. Listed: CvR-4.

Currants

BLACK NAPLES. WHITE GRAPE and RED DUTCH
Located: none. Listed (as above): CvR-4.

CHERRY
(1) "*Cherry/Currant*." (Watercolor; caption written in ink.) Located: APC. (A theorem and freehand version of this design was issued by D. M. Dewey c.1859: CvR-1, CvR-2.)
(2) "*Cherry Currant*." (A simplified version of #1.) Located: CU-1.
(3) "*The Cherry Currant*." Located: RU-1.
(4) (Same as #2?) Located: none. Listed: CUM-2.
(5) Located: none. Listed: CvR-4.

FERTILE DE ANGERS
Located: none. Listed: CvR-4.

RED GRAPE
Located: none. Listed: CvR-4.

VERSAILLES
Located: none. Listed: CvR-4.

VICTORIA
Located: none. Listed: CvR-4.

WHITE DUTCH. PRINCE ALBERT. BLACK NAPLES
(1) "*Currants*/I *White Dutch* II *Prince Albert* III *Black Naples*." (On one plate.) Located: CU-1, OSU.

(2) (Same as #1?) Located: none. Listed: CUM-2.

WHITE GRAPE
(1) "*White Grape Currant.*" Located: APC.
(2) "*White Grape Currant.*" (Caption written in pencil, details added to berries in watercolor.) Located: CvR-3.
(3) "*White Grape.*" ("Currant" added in pencil; does not have the details added to the berries as in #2). Located: CvR-2, CU-1, RU-2.
(4) Located: none. Listed: CvR-4.

Dewberries

MAMMOTH PROLIFIC
Located: none. Listed: CvR-4.

Gooseberries

AMERICAN SEEDLING
Located: none. Listed: CvR-4.

CROWN BOB
"*The Crown Bob Gooseberry./Melling's.*/Lith.& col[d.] by Amana Society, Homestead, Iowa." Located: APC.

CROWN BOB. WHITESMITH.
(1) "*Gooseberries. 1. Crown Bob. 2. Whitesmith.*" Located: CU-1.
(2) (Same as #1?) Located: APC.
(3) Located: none. Listed: CvR-4.

DOWNING'S
Located: none. Listed: CvR-4.

HOUGHTON'S SEEDLING
(1) "*Houghton's Seedling Gooseberry.*" Located: CU-1.
(2) (Same as #1?) Located: none. Listed: CUM-2.
(3) Located: none. Listed: CvR-4.

MOUNTAIN SEEDLING
Located: none. Listed: CvR-4.

SMITH'S IMPROVED
Located: none. Listed: CvR-4.

WHITESMITH
(1) "*The Whitesmith Gooseberry.* (Woodward's)/Lith. & col[d.] by Amana Society, Homestead Iowa." Located: AAS.
(2) (Same design as #1?); uncolored. Located: AMAH.

Grapes

ADIRONDAC
(1) "*The Adirondac Grape.*" Located: CU-1. Listed: CU-2.
(2) (Same as #1?) Located: AMAH.
(3) Located: none. Listed: CvR-4.

ALLEN'S HYBRID
(1) "*The Allen's Hybrid Grape.*" Located: CU-1. Listed: CU-2.
(2) (Same as #1?) Located: none. Listed: CUM-2, AAS-3.
(3) Located: none. Listed: CvR-4.

BRIGHTON
Located: none. Listed: CvR-4.

CALIFORNIA
Located: none. Listed: CvR-4.

CATAWBA
(1) "*The Catawba Grape.*/Painted by G. Prestele. Amana, Iowa." (Watercolor; caption and signature written in ink.) Located: AMAH.
(2) "*The Catawba Grape.*/Lith. & col[d.] by Amana Society. Amana, Iowa County, Iowa." (Revision of #1.) Located: CU-1.
(3) (Same as #2?) Located: none. Listed: CUM-2.
(4) Located: none. Listed: CvR-4.

CHASSELAS BLANC
(1) "*Chasselas Blanc Grape.*" Located: CU-1.
(2) Located: none. Listed: CUM-2.

CHRISTINE
Located: none. Listed: CvR-4.

CLINTON
(1) "*The Clinton Grape.*" Located: OSU.
(2) "*The Clinton Grape.*/Lith. by Amana Society. Homestead, Iowa." (Same design as #1; signature added.) Located: CU-1.
(3) (Same as #2?) Located: none. Listed: CUM-2.
(4) Located: none. Listed: CvR-4.

CONCORD
(1) "*The Concord Grape.*" Located: OSU.
(2) "*The Concord Grape.*" (Slight variation of #1.) Located: CU-1.
(3) (Same as #2?) Located: none. Listed: CUM-2.
(4) Located: none. Listed: CvR-4.

COPPER MINE
"*The Copper Mine Grape.*/Introduced by R. O. Thompson. Nursery HIll, Nebraska./Lith. & col[d.] by Amana Society. Amana, Iowa County, Iowa." Located: APC.

CREVELING
(1) "*The Creveling Grape.*/Introduced by R. O. Thompson, Nursery Hill, Nebraska./Lith. & col[d.] by Amana Society. Amana, Iowa County, Iowa." Located: CU-1. Listed: none.
(2) Located: none. Listed: CvR-4.

CROTON
Located: none. Listed: CvR-4.

DELAWARE
(1) "*Delaware Grape.*" Located: CU-1.
(2) (Same as #1?) Located: none. Listed: CUM-2.
(3) Located: none. Listed: CvR-4.

DIANA
(1) "*The Diana Grape.*" Located: CU-1.
(2) (Same as #1?) Located: none. Listed: CUM-2.

(3) Located: none. Listed: CvR-4.

EUMELAN
(1) "*The Eumelan Grape.*" Located: CU-1. Listed: none.
(2) (Same as #1?) Located: none. Listed: CUM-2.
(3) Located: none. Listed: CvR-4.

HARTFORD PROLIFIC
(1) "*The Hartford Prolific Grape.*" Located: CU-1.
(2) (Same as #1?) Located: none. Listed: CUM-2.

IONA
(1) "*The Iona Grape.*/Lith. & col[d.] by Amana Society. Iowa County, Iowa." Located: CU-1.
(2) (Same as #1?) Located: none. Listed: CUM-2.
(3)Located: none. Listed: CvR-4.

ISABELLA
(1) "*Isabella Grape.*" Located: CU-1, RU-2.
(2) "*Isabella.*" (Same as #1?) Located: none. Listed: CUM-2.
(3) "*Isabella.*" Located: none. Listed: CvR-4.

ISRAELLA
(1) "*The Israella Grape.*" Located: CU-1.
(2) "*Israella.*" Located: none. Listed: CUM-2.

IVES SEEDLING
Located: none. Listed: CvR-4.

MARTHA
(1) "*The Martha Grape.*" Located: CU-1. Listed: *not* on CU-2 list.
(2) (Same as #1?) Located: none. Listed: CUM-2.
(3) Located: none. Listed: CvR-4.

ONTARIO
Located: none. Listed: CvR-4.

REBECCA
(1) "*The Rebecca Grape.*" Located: CU-1.
(2) (Same as #1?) Located: none. Listed: CUM-2.
(3) Located: none. Listed: CvR-4.

ROGERS NO. 1. GOETHE
Located: none. Listed: CvR-4.

ROGERS NO. 3. MASSASOIT
Located: none. Listed: CvR-4.

ROGERS NO. 4. WILDER
(1) Located: none. Listed: CUM-2.
(2) Located: none. Listed: CvR-4.
(3) "*The Wilder Grape./Rogers No. 4.*" Located: CU-1.

ROGERS NO. 9. LINDLEY
Located: none. Listed: CvR-4.

ROGERS NO. 15. AGAWAM
Located: none. Listed: CvR-4.

ROGERS NO. 19. MERRIMACK
Located: none. Listed: CvR-4.

ROGERS NO. 43. BARRY
Located: none. Listed: CvR-4.

ROGERS NO. 53. SALEM
(1) Located: none. Listed: CvR-4.
(2) Located: none. Listed: CUM-2.

SCUPPERNONG
Located: none. Listed: CvR-4.

SENASQUA
Located: none. Listed: CvR-4.

THOMPSON'S WINE
"*The Thompson's Wine Grape.*/Introduced by R. O. Thompson, Nursery Hill, Nebraska./Lith. & col[d.] by Amana Society. Amana, Iowa County, Iowa." Located: RU-2.

WALTER
Located: none. Listed: CvR-4.

Mulberries

DOWNING'S EVER-BEARING MULBERRY
(1) "*Downing's Everbearing Mulberry.*" (Lithograph, engraved [?].) Located: CU-1. Listed (as "Dawnings Ever-Bearing Mulberry"): CU-2.
(2) (Same as #1?) Located: none. Listed (as "Dawnings Ever-Bearing Mulberry"): CUM-2.
(3) Located: none. Listed: CvR-4.

Nectarines

BOSTON
Located: none. Listed: CvR-4.

ELRUGE
Located: none. Listed: CvR-4.

GOLDEN
Located: none. Listed: CvR-4.

STANWICKS
Located: none. Listed: CvR-4.

Nuts

COMMON EUROPEAN WALNUT
(1) "*Common European Walnut.*" Located: CvR-2.
(2) (Same as #1?) Located: none. Listed: CU-2.
(3) (Same as #1?) Located: none. Listed: CUM-2.

CORYLLUS AVELLANA. HAZEL NUTS
(1) "*Coryllus Avellana Hazel Nut./I. Tubulosa. II. Alba. III Lambertiis. III [IV] d'Provence. V Colurna. VI d'Piedmont.*" Located: CU-1.
(2) (Same as #1?) Listed: CUM-2.

Peaches

ALBERGE YELLOW
(1) "*The Alberge Yellow Peach.* Lith. & col[d.] by Amana Society. Amana, Iowa County, Iowa." Located: AJM.
(2) "*The Alberge Yellow Peach.*" Located: RU-1.
(3) (Same as #1?) Located: none. Listed: CU-2, CUM-2.
(4) Located: none. Listed: CvR-4.

ALEXANDER'S EARLY
Located: none. Listed: CvR-4.

AMSDEN
Located: none. Listed: CvR-4.

BERGEN'S YELLOW
Located: none. Listed: CvR-4.

CHINESE CLING
Located: none. Listed: CvR-4.

COLUMBIA
Located: none. Listed: CvR-4.

COOPER'S MAMMOTH
Located: none. Listed: CvR-4.

CRAWFORD'S EARLY
(1) "*The Crawford's Early Peach.*" Located: CU-1.
(2) (Same as #1?) Located: none. Listed: CUM-2.
(3) Located: none. Listed: CvR-4.

CRAWFORD'S LATE
(1) "*Crawford's Late Peach.*" Located: CU-1.
(2) (Same as #1?) Located: none. Listed: CUM-2.
(3) Located: none. Listed: CvR-4.

EARLY ANN
Located: none. Listed: CvR-4.

EARLY BEATRICE
Located: none. Listed: CvR-4.

EARLY LOUISE
Located: none. Listed: CvR-4.

EARLY RIVERS
Located: none. Listed: CvR-4.

EARLY TILLOTSON
Located: none. Listed: CvR-4.

FOSTER
Located: none. Listed: CvR-4.

GEORGE THE FOURTH
Located: none. Listed: CvR-4.

HALE'S EARLY
(1) Located: none. Listed: CvR-4.
(2) Located: none. Listed: AAS-3.

HARPER'S CLING
Located: none. Listed: CvR-4.

HEATH CLING
Located: none. Listed: CvR-4.

HONEST JOHN
Located: none. Listed: CvR-4.

KENSINGTON
Located: none. Listed: CvR-4.

LARGE EARLY YORK
Located: none. Listed: CvR-4.

LARGE RARERIPE
(1) Located: none. Listed: CU-2.
(2) (Same as #1?) Located: none. Listed: CUM-2.
(3) Located: none. Listed (as "Large Red Rareripe"): CvR-4.

OLD MIXON FREE
Located: none. Listed: CvR-4.

PEACH APRICOT
(1) "*Apricot*" (Peach). Located: CU-1. Listed (as "Peach Apricot"): CU-2.
(2) (Same as #1?) Located: none. Listed: CUM-2.

POPE'S CLING
Located: none. Listed: CvR-4.

PRESIDENT
Located: none. Listed: CvR-4.

RED CHEEK MELACOTON
(1) "*Red Cheek Melacoton.*" Located: CU-1.
(2) (Same as #1?) Located: none. Listed: CUM-2.
(3) Located: none. Listed: CvR-4.

RED HEATH
Located: none. Listed: CvR-4.

REEVES' FAVORITE
Located: none. Listed: CvR-4.

SMOCK FREE
Located: none. Listed: CvR-4.

SNOW
(1) "*Snow Peach.*/Lith. & col[d] by Amana Society. Amana, Iowa County, Iowa." Located: CU-1.
(2) (Same as #1?) Located: none. Listed: CUM-2.

STUMP THE WORLD
(1) Located: none. Listed: CUM-2.
(2) Located: none. Listed: CvR-4.

SUSQUEHANNA
Located: none. Listed: CvR-4.

TROTH'S EARLY
Located: none. Listed: CvR-4.

WARD'S LATE
Located: none. Listed: CvR-4.

Pears

ANDREWS
(1) "*Andrews Pear.*" (Watercolor; caption written in ink. Beside the fruit is a penciled notation—"A little more yellow"—suggesting this was a proof.) Located: RU-2.
(2) "*The Andrews Pear.*" Located: RU-2, RU-1.
(3) "*The Andrews Pear.*" (Same as #2?) Located: CU-1.
(4) (Same as #2?) Located: none. Listed: CUM-2.

BARTLETT
(1) "*The Bartlett Pear.*" Located: OSU.
(2) "*The Bartlett Pear.*" (Printed from the same stone, and with the same caption, as #1.) Located: CUM-1.
(3) "*Bartlett.*" (Printed from the same stone as #1 but recaptioned.) Located: CU-1, CvR-2. (D. M. Dewey produced a theorem copy of this design c.1859, with the caption "Bartlett" and "D. M. Dewey. Rochester, N.Y.")
(4) "*Bartlett.*" (Same as #3?) Located: RU-1.

(5) "*Bartlett.*" (Same as #2?) (A simplified version of this design was used as an illustration in the *Horticulturist*, vol. 3, n.s. [August 1853], before p. 345. Both were produced by the same artist and, in the colored edition, with the same tinting.) Located: RU-2.
(6) "*Bartlett.*" (Same as #3?; uncolored.) Located: AAS.
(7) Located: none. Listed: CvR-4.

BELL OR KALLABASH
Located: none. Listed (as written addition to W. H. Prestele's 1879 catalogue): CvR-4.

BELLE LUCRATIVE
(1) "*The Belle Lucrative Pear.*" Located: CU-1.
(2) "*The Belle Lucrative Pear.*" (A slight revision of #1; the same lettered caption.) Located: CUM-1.
(3) Located: none. Listed: CvR-4.

BEURRE BROWN (See BROWN BEURRE, a synonym.)
(1) "*The Beurre Brown Pear.*" Located: CU-1.
(2) Located: none. Listed: CUM-2.

BEURRE CLAIRGEAU
(1) "*Beurre Clairgeau Pear.*" (Watercolor; caption written in ink.) Located: APC.
(2) "*The Beurre Clairgeau Pear.*" (A revision of #1.) Located: CU-1, CvR-2.
(3) "*The Beurre Clairgeau Pear.*" Located: RU-2.
(4) Located: none. Listed: CUM-2.

BEURRE D'ANJOU
(1) "*The Beurre D'Anjou Pear.*" Located: CvR-2.
(2) "*The Beurre D'Anjou Pear.*" (Same as #1?) Located: RU-2.
(3) "*The Beurre D'Anjou Pear.*" (Simplified version of #1.) Located: CUM-1. Listed: CUM-2.
(4) "*The Beurre D'Anjou Pear.*/Lith. & col[d.] by Amana Society. Amana, Iowa County, Iowa." (Same design as #1; Amana Society imprint added.)
(5) Located: none. Listed: CvR-4.
(6) "*The Beurre D'Anjou Pear.*" (Same as # 1, 2, and 3?) Located: AJM.

BEURRE D'AREMBERG
(1) "*The Beurre D'Aremberg Pear.*" Located: CU-1. Listed: CU-2.
(2) Located: none. Listed: CUM-2.
(3) "*The Beurre D'Aremberg Pear.*/Lith. & col[d.] by Amana Society. Amana, Iowa County, Iowa." (Printed from the same stone as #1; Amana Society imprint added.) Located: RU-2. (Plates [1] and [3] are copies, with minor additions, of Joseph Prestele's unsigned "Beurre D'Aremberg" plate in the *Transactions of The Massachusetts Horticultural Society*, vol. 1, no. 3 [January]1852).

BEURRE DE WATERLOO
(1) "*Beurre De Waterloo.*" Located: RU-2.
(2) "*Beurre De Waterloo.*" Located: CU-1.
(3) Located: none. Listed: CUM-2.

BEURRE DIEL
(1) "*The Beurre Diel Pear.*/Lith. & col[d.] by Amana Society. Amana, Iowa County, Iowa." Located: CU-1, RU-2.
(2) (Same as #1?) Located: none. Listed: CUM-2.
(3) Located: none. Listed: CvR-4.

BEURRE GIFFARD
(1) "*The Beurre Giffard Pear.*" Located: CvR-2.
(2) (Same as #1?) Located: RU-2.
(3) "*The Beurre Giffard Pear.*/Lith. & col[d.] by Amana Society. Amana, Iowa County, Iowa." (Revision of #1.) Located: CU-1. (A different version of this subject by Joseph Prestele was used as an illustration in the *Horticulturist*, vol. 4, n.s. [February 1854].)
(4) Located: none. Listed: CUM-2.
(5) Located: none. Listed: CvR-4.

BEURRE HARDY
(1) "*Beurre Hardy.*" Located: CU-1.
(2) (Same as #1?) Located: none. Listed: CUM-2.
(3) "*Beurre Hardy.*" (Same as # 1 and 2?) Located: AJM.

BEURRE SUPERFIN
(1) "*The Beurre Superfin Pear.*" Located: CU-1. (This plate is identical to the illustration in the *Horticulturist*, vol. 7, n.s. [January 1857], including the caption lettering.)
(2) Located: none. Listed: CvR-2.

BLOODGOOD
(1) "*The Bloodgood Pear.*" Located: CU-1, CUM-1.
(2) Located: none. Listed: CvR-4.

BRANDYWINE
(1) "*The Brandywine Pear.*" Located: CvR-2, CU-1.
(2) "*The Brandywine Pear.*" (Same as #1?) Located: RU-2.
(3) (Same as #1?) Located: none. Listed: CUM-2.
(4) Located: none. Listed: CvR-4.

BROWN BEURRE (See BEURRE BROWN, a synonym.)
"*The Brown Beurre Pear.*/by J. & G. Prestele. Amana, Iowa Co., Iowa." (Watercolor; caption and signature written in ink. "The Beurre Brown Pear" is a modified version of this design.) Located: RU-2.

BUFFUM
(1) "*The Buffum Pear.*" Located: RU-1.
(2) "*The Buffum Pear.*/Lith. & col[d.] by Amana Society, Amana, Iowa County, Iowa." Located: CU-1.
(3) (Same as #2?) Located: none. Listed: CUM-2.
(4) Located: none. Listed: CvR-4.
(5) "*The Buffum Pear.*" (Uncolored.) Located: AJM.

CALEBASSE MONSTREUSE
Located: none. Listed: CvR-4.

CATHARINE
Located: none. Listed (as written addition to W. H. Prestele's 1879 catalogue): CvR-4.

CHRISTIAN
"*The Christian Pear.*" Located: APC.

CLAPP'S FAVORITE
Located: none. Listed: CvR-4.

DEARBORN'S SEEDLING
Located: none. Listed: CvR-4.

DISC [DIX?]
Located: none. Listed: CvR-4.

DOYENNE BOUSSOCK [BOISSELOT?]
"*The Doyenne Boussock Pear.*/Lith. & col[d.] by Amana Society. Amana, Iowa County, Iowa." Located: CU-1, CUM-1.

DOYENNE D'ALENCON See: DOYENNE D'HIVER D'ALENCON

DOYENNE D'ETE
(1) "*The Doyenne D'Ete Pear.*" Located: CU-1, CUM-1.
(2) Located: none. Listed: CvR-4.

DOYENNE D'HIVER D'ALENCON
(1) "*The Doyenne D'Hiver D'Alencon Pear.*" Located: RU-1.
(2) (Same as #1?) Located: CU-1, CUM-1. Listed (as "Doyenne d'Alencon"): CU-2, CUM-2.
(3) (Same as #1?) Located (uncolored): AMAH.

DOYENNE WHITE or VIRGALIEU—See VIRGALIEU or WHITE DOYENNE

DUCHESSE D'ANGOULEME
(1) "*The Duchesse D'Angouleme Pear.*" Located: CU-1, CUM-1.
(2) Located: none. Listed (as "Dutchess de Angouleme"): CvR-4.

DUCHESSE DE BERRY D'ETE
(1) "*The Duchesse De Berry D'Ete Pear.*/G. Prestele." (A variant design of the "Madeline" pear, see entry.) Located: CU-1, RU-1, RU-2.
(2) (Same as #1?) Located (uncolored): AMAH.
(3) Located: none. Listed: CvR-4.

DUCHESSE D'ORLEANS
(1) "*Duchesse D'Orleans Pear.*" Located: RU-2, CU-1. Listed: CU-2. (A variation of the Joseph Prestele plate in the *Horticulturist*, vol. 3, n.s. [July 1853].)
(2) (Same as #1?) Located: none. Listed: CUM-2.
(3) Located: none. Listed (as "Dutchess de Orleans"): CvR-4.

EASTER BEURRE
(1) "*Easter Beurre Pear.*" Located: CvR-2, RU-1.
(2) "*The Easter Beurre Pear.*/Lith. & col[d.] by Amana Society. Amana, Iowa County, Iowa." Located: CU-1. Listed: CU-2. (A slightly altered version of the Joseph Prestele plate in the *Horticulturist*, vol. 4, n.s. [October 1854].)
(3) (Same as #2?) Located: none. Listed: CUM-2.

EDMONDS
Located: none. Listed: CvR-4.

FLARE (?) (Not in Ragan, *Nomenclature of the Pear.*) Located: none. Listed (as written addition to W. H. Prestele's 1879 catalogue): CvR-4.

FLEMISH BEAUTY
(1) "*The Flemish Beauty Pear.*" Located: OSU, CU-1, CUM-1.
(2) Located: none. Listed: CvR-4 (and colored frontispiece).
(3) "*Flemish Beauty*/Lith. & col[d.] by W. H. Prestele, Iowa City, Iowa." Located: APC.

GLOUT MORCEAU
(1) "*The Glout Morceau Pear.*" Located: RU-1.
(2) (Same as #1?) Located: none. Listed: CUM-2.
(3) "*The Glout Morceau Pear.*" (Same as #1?) Located: CU-1.

GRAY DOYENNE
(1) "*The Gray Doyenne Pear.*" (Watercolor; caption written in ink.) Located: APC.
(2) "*The Gray Doyenne Pear.*" Located: RU-1, RU-2.
(3) (Same as #2?) Located: none. Listed: CU-2, CUM-2.
(4) Located: none. Listed: CvR-4.
(5) "*The Gray Doyenne Pear.*" (Uncolored) Located: AJM.

HOWELL
(1) "*Howell Pear.*" Located: CU-1. (A simplified version of the Joseph Prestele plate in the *Horticulturist*, vol. 5, n.s. [August 1855].)
(2) (Same as #1?) Located: none. Listed: CUM-2.
(3) (Same as #1?) Located: RU-2.
(4) Located: none. Listed: CvR-4.

LAWRENCE
(1) "*The Lawrence Pear.*" Located: CU-1.
(2) Located: none. Listed: CUM-2.

LOUISE BONNE DE JERSEY
(1) "*The Louise Bonne De Jersey Pear.*" Located: CvR-2.
(2) "*The Louise Bonne De Jersey Pear.*" (Same as #1?) Located: RU-2, CU-1.
(3) "*The Louise Bonne De Jersey Pear.*" (Revision of #2.) Located: CUM-1.
(4) Located: none. Listed (as "Louise Bon De Jersey"): CvR-4.

MADELEINE
(1) "*Madelaine Pear.*/G. Prestele." (A slightly reworked version of the "Duchesse De Berry D'Ete Pear" plate.) Located: OSU.
(2) "*Madeline.*/G. Prestele." (Slight revision of #1, including lettering.) Located: RU-2.
(3) (Same as #2?) Located: none. Listed: CUM-2.
(4) Located: none. Listed (as "Madeline"): CvR-4.

MARIE LOUISE
Located: none. Listed: CvR-4.

MT. VERNON
Located: none. Listed: CUM-2.

NAPOLEON
(1) "*The Napoleon Pear.*" (Watercolor; caption

written in ink.) Located: RU-2.
(2) "*The Napoleon Pear.*" Located: CU-1.
(3) Located: none. Listed: CUM-2.

NOUVEAU POITEAU
(1) "*The Nouveau Poiteau Pear.*" Located: CU-1. Listed: CU-2.
(2) Located: none. Listed: CUM-2.

OATS (?) (Not in Ragan, *Nomenclature of the Pear.*) Located: none. Listed (as written addition to W. H. Prestele's 1879 catalogue): CvR-4.

OSBAND'S SUMMER
(1) "*Osband's Summer Pear.*" Located: CU-1. (A simplified version of the Joseph Prestele plate in the *Horticulturist*, vol. 5, n.s. [April 1855].)
(2) "*Osband's Summer Pear.*" (Same as #1?) Located: RU-2.
(3) Located: none. Listed: CUM-2.
(4) Located: none. Listed: CvR-4.

OSWEGO BEURRE
(1) "*The Oswego Beurre Pear*/Lith. & col$^{d.}$ by Amana Society. Amana, Iowa County, Iowa." Located: CU-1.
(2) "*The Oswego Beurre Pear.*" Located: RU-2.
(3) Located: none. Listed: CUM-2.
(4) Located: none. Listed: CvR-4.

PARADISE D'AUTOMNE PEAR
(1) "*The Paradise D'Automne Pear.*/By J. & G. Prestele. Amana. Iowa Co. Iowa." (Watercolor; caption and signature written in ink.) Located: RU-2.
(2) "*The Paradise D'Automne Pear.*" (Extensive revision of the original plate, #1.) Located: CU-1.
(3) (Same as #2?) Located: none. Listed: CUM-2.
(4) Located: none. Listed: CvR-4.

PRATT
(1) "*The Pratt Pear*/By J. & G. Prestele. Amana Iowa Co. Iowa." (Watercolor; caption and signature written in ink.) Located: RU-2.
(2) "*The Pratt Pear.*" (A revision of the original design, #1.) Located: CU-1.
(3) (Same as #2?) Located: none. Listed: CUM-2.
(4) Located: none. Listed: CvR-4.

RAPELJE
"*The Rapaljie Pear.*/By J. & G. Prestele. Amana Society, Iowa." (Watercolor; caption and signature written in ink.) Located: RU-2.

ROSTIEZER
(1) "*Rostiezer Pear.*" Located: RU-2.
(2) "*Rostiezer Pear.*" Located: CU-1. Listed: CU-2. (A variation of the Joseph Prestele plate in the *Horticulturist*, vol. 4, n.s. [August 1854].)
(3) (Same as #2?) Located: none. Listed: CUM-2.
(4) Located: none. Listed: CvR-4.

SECKEL
(1) "*Seckel Pear.*" (Watercolor; caption written in ink.) Located: APC.
(2) "*The Seckel Pear.*" (Simplified version of original, #1.) Located: APC.
(3) "*The Seckel Pear.*/Lith. & col$^{d.}$ by Amana Society, Amana, Iowa County, Iowa." (Variation of #2.) Located: CU-1. Listed: CU-2.
(4) "*The Seckel Pear.*" (A redrawn, simplified version of #3.) Located: CUM-1. Listed: CUM-2.
(5) Located: none. Listed: CvR-4.

SHELDON
(1) "*Sheldon Pear.*" (Lithograph; engraved. Small plate, 24.5 × 17.5 cm. Lightly tinted.) Located: CvR-1. (A simplified version of this design was used as the frontispiece to the *Horticulturist*, vol. 3, n.s. [January 1853].)
(2) "*The Sheldon Pear*, Lith. & col$^{d.}$ by Amana Society. Amana, Iowa County, Iowa." Located: CU-1, CUM-1.
(3) Located: none. Listed: CvR-4.
(4) "Sheldon Pear?" (Watercolor. Attributed to W. H. Prestele.) Located: AMAH.

SOUVENIR DU CONGRESS
Located: none. Listed: CvR-4.

STERLING
(1) "*Sterling Pear.*" (Watercolor; caption written in ink.) Located: APC.
(2) "*The Sterling Pear.*" (Exact reproduction of the original design, #1.) Located: CU-1.
(3) (Same as #2?) Located: RU-1, RU-2.
(4) Located: none. Listed: CUM-2.
(5) Located: none. Listed: CvR-4.

STEVENS GENESEE
(1) "*The Stevens Genesee Pear.*"/By J. Prestele. Amana Society. Iowa Co. Iowa." (Watercolor; caption and signature written in ink.) Located: RU-2.
(2) "*Stevens' Genesee.*/Lith. & col$^{d.}$ by Amana Society. Amana, Iowa County, Iowa." Located: CU-1.
(3) (Same as #2?) Located: none. CUM-2.
(4) Located: none. Listed: CvR-4.
(5) "*The Stevens' Genesee Pear*" (Uncolored) Located: AMAH.

SWAN'S ORANGE
(1) "*The Swan's Orange Pear.*" Located: RU-1, CvR-2.
(2) "*Swan's Orange Pear.*" Located: CU-1.
(3) (Same as #2?) Located: none. Listed: CUM-2.
(4) Located: none. Listed: CvR-4.
(5) (Same as # 2 and 3?) Located: none. Listed: AAS-3.

TYSON
(1) "*Tyson Pear.*" Located: RU-2. (A simplified version of this plate by Joseph Prestele is in the *Horticulturist*, vol. 5, n.s. [February 1855].)
(2) "*The Tyson Pear.*/Kelloggs Series of Fruits, Flowers and Ornamental Trees./Lith. of E. B. & E. C. Kellogg. Hartford, Conn./28." Ca. 1858. Design attributed to Joseph Prestele. Located: CvR-3.
(3) "*The Tyson Pear.*" Located: AMAH.
(4) "*Tyson.*" Located: none. Listed: CU-2, CUM-2.
(5) "*Tyson.*" Located: none. Listed: CvR-4.

VAN ASSCHE
(1) "*Vanassch Pear.*" (Watercolor; caption written in ink.) Located: RU-1.
(2) "*The Van Assche Pear.*" Located: RU-1, RU-2.
(3) "*The Van Assche Pear.*" Located: CU-1.

VAN MONS LEON LE CLERC
(1) "*The Van Mons. Leon Le Clerc Pear.*/By J. & G. Prestele. Amana Society, Iowa." (Watercolor; caption written in ink.) Located: RU-2.
(2) "*The Van Mons Leon Le Clerc Pear.*/Lith. & col^d. by Amana Society. Amana, Iowa County, Iowa." Located: CU-1.
(3) (Same as #1?) (Uncolored) Located: AJM.
(4) (Same as #2?) Located: none. Listed: CUM-2.
(5) Located: none. Listed (as "Van Mons Leonie Clerc"): CvR-4.

VICAR OF WINKFIELD
(1) "*Vicar of Winkfield.*" Located: CvR-2.
(2) "*Vicar of Winkfield.*" Located: RU-1.
(3) "*The Vicar of Winkfield Pear.*/Lith. & col^d. by Amana Society. Amana, Iowa County, Iowa." Located: CU-1.
(4) Located: none. Listed: CUM-2.

VIRGALIEU OF WHITE DOYENNE
(1) "*The Virgalieu or White Doyenne Pear.*" Located: CvR-2.
(2) "*The Virgalieu or White Doyenne Pear.*/Lith. & col^d. by Amana Society. Amana, Iowa County, Iowa." Located: RU-2.
(3) "*The Virgalieu or White Doyenne Pear.*/Lith. & col^d. by Amana Society. Amana, Iowa County, Iowa." Located: CU-1, CUM-1.
(4) Located: none. Listed: CvR-4.

WATER MELON
Located: none. Listed (as addition to W. H. Prestele's 1879 catalogue): CvR-4.

WHITE DOYENNE See: VIRGALIEU OF WHITE DOYENNE

WINTER NELIS
"*Winter Nelis.*/Lith. & col^d. by Amana Society. Amana, Iowa County, Iowa." Located: AJM.

Plums

BRADSHAW
(1) "*The Bradshaw Plum.*" (Watercolor; caption written in ink.) Located: APC.
(2) "*Bradshaw Plum.*" (Simplified revision of #1.) Located: CU-1, CvR-2.
(3) "*The Bradshaw Plum.*" (Same as #2?) Located: RU-1.
(4) Located: none. Listed: CUM-2.
(5) Located: none. Listed: CvR-4.

COE'S GOLDEN DROP
(1) Located: none. Listed: CU-1, CUM-2.
(2) Located: none. Listed: CvR-4.

DAMSON
Located: none. Listed: CvR-4.

DENNISON'S SUPERB
(1) "*The Dennison's Superb Plum.*" Located: RU-1.
(2) "*The Dennison's Superb Plum.*" Located: CU-1, CvR-2.
(3) (Same as #2?) Located: none. Listed: CUM-2.
(4) Located: none. Listed: CvR-4.

DESSERT
Located: none. Listed (as written addition to 1879 catalogue): CvR-4.

DUANE'S PURPLE
(1) "*Duane's Purple.*" Located: CvR-2. (Almost identical to the Joseph Prestele plate in the *Horticulturist*, vol. 5, n.s. [June 1855]. #3 was printed in reverse of that design.)
(2) (Same as #1?) Located: RU-2.
(3) "*The Duane's Purple Plum.*" (Simplified version of #1.) Located: CU-1.
(4) (Same as #3?) Located: none. Listed: CUM-2.
(5) Located: none. Listed: CvR-4.

GENERAL HAND
Located: none. Listed: CvR-4.

GERMAN PRUNE
Located: none. Listed: CvR-4.

GREEN GAGE OF REINE CLAUDE
(1) "*The Green Gage Plum or (Reine Claude.)*/Lith. & col^d. by Amana Society. Amana, Iowa County, Iowa." Located: CU-1.
(2) (Same as #1?) Located: none. Listed: CUM-2.
(3) Located: none. Listed: CvR-4.

IMPERIAL GAGE
Located: none. Listed: CvR-4.

JEFFERSON
(1) (Large, horizontal design; three plums.) "*Jefferson Plum.*" Located: CU-1.
(2) (Vertical design created from above; two plums.) "*Jefferson.*" (Uncolored.) Located: AMAH.
(3) (Original design reduced to one plum.) "*The Jefferson Plum.*" (Uncolored.) Gottlieb Prestele's file copy with a penciled notation that 60 copies were printed January 11, 1870, and 150 more on December 26, 1875. Located: AMAH.
(4) "*Jefferson*/Drawn from Nature Lith & col^d. by W. H. Prestele, Iowa City, Iowa." Located: AES.

LAWRENCE'S FAVORITE
(1) "*The Lawrence's Favorite Plum.*" Located: RU-1, CU-1.
(2) (Same as #1?) Located: none. Listed: CUM-2.

LOMBARD
(1) Located: none. Listed: CUM-2.
(2) Located: none. Listed: CvR-4.

MCLAUGHLIN
(1) "*McLaughlin Plum.*" (Watercolor; caption written in ink.) Located: APC.
(2) "*The Mc Laughlin Plum.*/J. Prestele. Ebenezer, n. Buffalo." (A variant of #1.) Located: CU-1, RU-1.

(3) Located: none. Listed: CvR-4.

MAGNUM BONUM
Located: none. Listed: CvR-4.

MINER
(1) "*The Miner Plum.*" Located: APC.
(2) Located: none. Listed: CvR-4.

MIRABELLA
"*Mirabella Plum.*" (Watercolor; caption written in ink.) Located: APC.

NEBRASKA SEEDLING. THOMPSON'S GOLDEN GEM
"1. *The Nebraska Seedling Plum.* 2. *Thompson's Golden Gem Plum.*/Introduced by R. O. Thompson, Nursery Hill, Nebraska./Lith. & col[d.] by Amana Society. Amana Iowa County, Iowa." Located: RU-2.

PEACH
(1) "*Peach Plum.*" (Watercolor; caption written in ink. Attributed by AMAH, perhaps incorrectly, to W. H. Prestele. Printed from stone in AMAH collection.) Located: AMAH.
(2) "*The Peach Plum.*" (Slight modification of #1.) Located: CU-1.
(3) "*The Peach Plum.*" (Uncolored; printed from stone in AMAH collection.) Located: AMAH.
(4) "*The Peach Plum.*" (Same as #2?) Located: RU-1.
(5) Located: none. Listed: CvR-4.

POND'S SEEDLING
(1) "*The Pond's Seedling Plum.*" Located: RU-1.
(2) "*The Pond's Seedling Plum.*" Located: CvR-2.
(3) "*The Pond's Seedling Plum.*" (Simplified version of #2.) Located: CU-1.
(4) Located: none. Listed: CUM-2.
(5) Located: none. Listed: CvR-4.

PRUNE D'AGEN
(1) "*Prune D'Agen Plum.*" (Watercolor; caption written in ink.) Located: RU-2.
(2) "*The Prune D'Agen or Robe De Servent Plum.*" (A completely revised and more elaborate design than #1.) Located: CU-1, RU-1, RU-2.
(3) Located: none. Listed: CUM-2.

RICHLAND
Located: none. Listed: CvR-4.

SMITH'S ORLEANS
Located: none. Listed: CvR-4.

WASHINGTON
(1) "*The Washington Plum.*" Located: CU-1.
(2) (Same as #1?) Located: none. Listed: CUM-2.

WEAVER
Located: none. Listed: CvR-4.

WILD GOOSE
Located: none. Listed: CvR-4.

YELLOW EGG
Located: none. Listed: CvR-4.

Quinces

ANGERS
Located: none. Listed: CvR-4.

APPLE OF ORANGE (See ORANGE QUINCE)

ORANGE QUINCE
(1) "*Orange Quince.*" Located: CU-1. Listed: CU-2.
(2) "*Orange Quince.*" (Uncolored: Gottlieb Prestele's file copy with penciled notation that 25 copies were printed on April 2, 1863, 50 on January 19, 1864, 50 on January 9, 1865, 50 on February 21, 1866, 30 on December 1, 1869, 200 on January 11 and 12, 1871. Located: AMAH.
(3) Located: none. Listed (as "Quince"): CUM-2.
(4) Located: none. Listed (as "Apple or Orange"): CvR-4.

REA'S MAMMOTH
Located: none. Listed: CvR-4.

Raspberries

CLARKE
Located: none. Listed (as "Clarks"): CvR-4.

DAVISON'S THORNLESS
Located: none. Listed: CvR-4.

DOOLITTLE'S BLACK CAP
(1) "*Doolittles Black Cap Raspberry.* Lith. & col[d.] by Amana Society, Iowa." (Uncolored.) Located: AS.
(2) Located: none. Listed: CUM-2.
(3) Located: none. Listed: CvR-4.

ELLISDALE
(1) "I. *Elsdale Raspberry.* 2. *The Nebraska Prolific Gooseberry.*/Introduced by R. O. Thompson, Nursery Hill, Nebraska./Lith. & col[d.] by Amana Society. Amana, Iowa County, Iowa." Located: RU-2.
(2) "I. *Elisdale Raspberry.* 2. *The Downings Seedling Gooseberry.*/Introduced by R. O. Thompson, Nursery Hill, Nebraska/Lith. & col[d.] by Amana Society, Amana, Iowa County, Iowa." (Variant of #1 with spelling of "Elsdale" corrected, and "Downings Seedling" substituted for "Nebraska Prolific." These corrections were made on the stone used for printing #1.) Located: CU-1.
(3) "1. *Elisdale Raspberry.* 2. *The American Seedling Gooseberry.*/Introduced by R. O. Thompson, Nursery Hill, Nebraska./Lith. & col[d.] by Amana Society. Amana, Iowa County, Iowa." Located: RU-2.

GOLDEN CAP
Located: none. Listed: CvR-4.

GOLDEN THORNLESS
Located: none. Listed: CvR-4.

GREGG
Located: none. Listed (as written addition to catalogue): CvR-4.

HERSTINE
Located: none. Listed (as "Herstien"): CvR-4.

MAMMOTH CLUSTER
(1) Located: none. Listed: CUM-2.
(2) Located: none. Listed: CvR-4.

MIAMI
Located: none. Listed: CvR-4.

OHIO EVERLASTING
Located: none. Listed: CvR-4.

PHILADELPHIA
Located: none. Listed: CvR-4.

PURPLE CANE
Located: none. Listed: CvR-4.

RED ANTWERP
(1) "*The Red Antwerp Raspberry*." Located: APC.
(2) "*Raspberries/I Red Antwerp*. II *Yellow or White./* Lith. & col[d.] by Amana Society. Amana, Iowa County, Iowa." Located: CU-1.
(2) (Same as #?) Located: none. Listed: CUM-2.

TURNER
Located: none. Listed (as written addition to W. H. Prestele's 1879 catalogue): CvR-4.

YELLOW RASPBERRY
"*The Yellow Raspberry*." Located: AAS. Listed (as "Yellow or White" Raspberry): AAS-3.

Strawberries

AGRICULTURIST
(1) "*The Agriculturist Strawberry*." Located: CU-1. Listed: CU-2. (The lithographic stone used in printing this plate is at the Amana Heritage Society Museum.)
(2) (Same as #1?) Located: none. Listed: CUM-2.
(3) Located: none. Listed: CvR-4.

AUSTIN SHAKER
(1) "*Austin Shaker Strawberry*." Located: CU-1.
(2) (Same as #1?) Located: none. Listed: CUM-2.

BICTON PINE
(1) "*Bicton Pine./Strawberry*." (Watercolor; caption written in ink.) Located: APC.
(2) "*Bicton Pine*." (A simplified version of #1.) Located: CU-1, RU-1, RU-2.
(3) (Same as #2?) Located: none. Listed: CUM-2.

CAPT. JACK
Located: none. Listed: CvR-4.

CHARLES DOWNING
Located: none. Listed: CvR-4.

COLONEL CHENEY
Located: none. Listed: CvR-4.

CRESCENT SEEDLING
Located: none. Listed (as written addition to 1879 catalogue): CvR-4.

DOWNER'S PROLIFIC
Located: none. Listed: CvR-4.

DR. NICAISE
Located: none. Listed: CvR-4.

GENESEE SEEDLING
(1) "*Genesee Seedling*." Located: RU-2, RU-1, CU-1.
(2) (Same as #1?) Located: none. Listed: CUM-2.

GREAT AMERICAN
Located: none. Listed: CvR-4.

GREEN PROLIFIC
(1) "*The Green Prolific Strawberry*." (A reissue of "The Monroe Scarlet Strawberry" plate with "Green Prolific" substituted for "Monroe Scarlet" in the caption; see MONROE SCARLET.) Located: OSU.
(2) Located: none. Listed: CvR-4.

HONNEUR DE BELGIQUE
(1) "*Honneur De Belgique*." Located: CU-1. (A theorem and freehand simplified version of this design was published c.1859 with the printed caption "Honneur de Belgique" and the imprint: "D. M. Dewey, Rochester, N.Y.")
(2) (Same as #1?) Located: RU-2.

HOOKERS SEEDLING
(1) "*Hookers Seedling*." (Caption written in pencil.) Located: CvR-3.
(2) "*Hookers Strawberry*." (A simplified version of #1.) Located: CU-1.
(3) (Same as #2?) Located: none. Listed: CUM-2.
(4) Located: none. Listed (as "Hooker"): CvR-4.

HOVEY'S SEEDLING
(1) "*Hoveys' Seedling*." Located: RU-2, CU-1. Listed: CU-2. (D. M. Dewey issued an almost exact copy of #1 in theorem and freehand, c.1859, with a printed caption and "D. M. Dewey, Rochester, N.Y.") Listed: CvR-3.
(2) (Same as #1?) Located: none. Listed: CUM-2.

JUCUNDA
Located: none. Listed: CvR-4.

KENTUCKY
Located: none. Listed: CvR-4.

MCAVOY'S SUPERIOR
(1) "*McAvoy's Superior Strawberry*." Located: CU-1. (Adapted from the Joseph Prestele plate in the *Horticulturist*, vol. 3, n.s. [September 1853].)
(2) "*McAvoy's Superior*." Located: RU-2.
(3) (Same as #1?) Located: none. Listed: CUM-2.

MONARCH OF THE WEST
Located: none. Listed: CvR-4.

MONITOR
Located: none. Listed: CvR-4.

MONROE SCARLET
(1) "*The Monroe Scarlet Strawberry*." Located: RU-1.
(2) (Same as #1?) Located: CU-1.
(3) (Same as #1?) Located: none. Listed: CUM-2.

RED JACKET
Located: none. Listed (as written addition to 1879 catalogue): CvR-4.

RUSSELL'S GREAT PROLIFIC
Located: none. Listed: CvR-4.

SHARPLESS SEEDLING
Located: none. Listed: CvR-4.

STAR OF THE WEST
Located: none. Listed: CvR-4.

STRAWBERRIES. Ohio Mammoth. Burr's New Pine. McAvoy's Superior. Black Prince. Triomphe de Gand.
"*Strawberries*/I *Ohio Mammoth*. II *Burr's New Pine*. III *McAvoy's Superior*. IV *Black Prince*. V *Triomphe de Gand*." Located: RU-1, RU-2, CU-1.

TRIOMPHE DE GAND
(1) "*Triomphe De Gand*." Located: CvR-2, CU-1, RU-1. (The lithographic stone used for printing this plate is at the Museum of Amana History.
(2) (Same as #1?) Located: none.
(3) Located: none. Listed: CvR-4.

TROLLOPE'S VICTORIA
(1) "*Trollopes Victoria*." Located: CvR-2, CU-1, RU-1, RU-2.
(2) (Same as #1?) Located: none. Listed: CUM-2.

WILSON'S ALBANY
(1) "*The Wilson Strawberry*." (Watercolor; caption written in ink.) Located: APC.
(2) "*The Wilson's Albany Strawberry*." Located: CU-1.
(3) "*The Wilson's Albany Strawberry*." Located: AMAH.
(4) "*Wilsons Albany Strawberry*." (Completely different design from #2.) Located: OSU.
(5) (Same as #2?) Located: none. Listed (as "Wilson's Albany Seedling"): CU-2, CUM-2.
(6) Located: none. Listed (as "Wilson's Albany"): CvR-4.

Flowers

BOUQUET OF FLOWERS, tied with ribbon
(This may not have been made as a nurserymen's plate but as an illustration or a decorative piece for framing. Not listed.)
(1) (Uncolored.) Located: AMAH.
(2) (Colored): APS.

"CHINESE POEONIAS [*sic*]. 1. P. Potsii. 3. P. Carnea. 3. P. Compte de Paris."
(1) Uncolored. Located: AJM.
(2) Colored. Located: CvR-1. Not listed.

"CHRYSANTHEMUMS"
(1) Located: none. Listed: CU-2, CUM-2.
(2) Located: none. Listed: CvR-4.

"CHRYSANTHEMUMS. 1 La Gitano 2 Criterion 3 Asmodea 4 Daphnis 5 Perfecta."
(The same as the "Chrysanthemums" plate above?) Located: RU-2.

"CROCUS (three specimens)."
(1) Located: none. Listed: CU-2, CUM-2.
(2) Located: none. Listed: CvR-4.

"DAHLIA"
(1) "*Dahlia*." Located: private collection.
(2) (Same as #1?) Listed: CU-2, CUM-2.
(3) Located: none. Listed: CvR-4.

"DIELYTRA SPECTABILIS"
(1) Located: APC. (Design used c. 1859 in a theorem and freehand plate with a printed caption and "D. M. Dewey, Rochester, N.Y." imprint. Both plates related in design to the Joseph Prestele illustration, "Dicentra or Dielytra Spectabilis," in the *Horticulturist*, vol. 4, n.s. [July 1854].)
(2) Located: none. Listed: CU-2, CUM-2, AAS-3.
(3) Located: none. Listed (as "Dielytra Spectabilis, or Bleeding Heart"): CvR-4.
(4) "*Dielytra Spectabilis*"/Lith. & col^d. by W. H. Prestele, Iowa City, Ia." Located: AES.
A lithographic stone with a "Dielytra Spectabilis" image and caption is at AMAH. It was used in printing #1 and/or #2 above.

"DOUBLE FLOWERING SWEET VIOLETS and SNOW DROPS."
(1) I *Double Flowering Sweet Violets* II *Snow Drops*." Located: AES.
(2) (Same as #1?) Located: none. Listed: CU-2, CUM-2.
(3) Located: none. Listed: CvR-4.

"FUCHSIA"
Located: none. Listed: CvR-4.

"GLADIOLUS"
(1) Located: none. Listed: CU-2, CUM-2, AAS-3.
(2) Located: none. Listed: CvR-4.
(3) (Same design as #1?; uncolored.) Located: AS.

"HEMEROCALIS FLAVA"
(1) Located: none. Listed (as above): CU-2, CUM-2, AAS-3.
(2) Located: none. Listed (as "Hemerocallis Flava"): CvR-4.

"HYACINTH"
(1) Located: none. Listed: CU-2, CUM-2.
(2) Located: none. Listed: CvR-4.

IRIS
Located: none. Listed: CU-2, CUM-2.

LILIUM AURATUM
Located: none. Listed: CvR-4.

LILIUM LANCIFOLIUM RUBRUM
(1) Located: CvR-2.
(2) (Same as #1?) Located: RU-1.
(3) Located: none. Listed (as "Lilium L. Rubrum"): CvR-4.

PAEONY HUMEI
Located: none. Listed: CvR-4.

"PANSIES"
"Gemalt von G. Prestele. Amana, Iowa." (Watercolor; caption and signature written in ink.) Located: APC.

"PETUNIAS"
"Executed by J. Prestele. Amana Iowa" (Water-

color; caption and signature written in ink.) Located: APC.

TULIP
(1). "*Tulip.*" (Uncolored.) Located: AS. (See TULIPS below.)
(2). "*Double Tulip.*" Same as #1? Located: none. Listed: CvR-4.

TULIPS
(1). "*Tulips.*" Located: CvR-1.
The design was adapted from the *Tulip* plate listed above. It consists of the double red tulip (printed in reverse), and the addition of a single "bizarre" tulip. Also, the caption was revised by the addition of a final "s."
(2). "*Tulips.*" Same as #1? Listed: CU-2, CUM-2.

TRITOMA
Located: none. Listed: CvR-4.

"VERBENAS"
Located: none. Listed: CvR-4.

Roses

AGRIPPINA
Located: none. Listed (as above) CvR-4.

AUGUSTA
(1) "*The Augusta Rose.*" Located: AMAH.
(2) "*The Augusta Rose.*" (Same as #1?) Located: RU-2.
(3) "*Augusta*/D. M. Dewey, Rochester, N.Y." Located: CvR-3.
(4) (Same as #1?) Located: none. Listed: CU-2, CUM-2.
(5) "*Augusta*, or Solfaterre." Located: none. Listed: CvR-4.

AUGUSTE MIE
(1) "*Augusta Mie.*" (Watercolor? Handwritten caption.) Located: APC.
(2) "*Augusta Mie.* Amana Society by I. & G. Prestele. Amana Iowa County Iowa." (Same design as #1? Only the leaves and stem colored; proof copy?) Located: AMAH.
(3) Located: none. Listed (as "Augusta Mie"): CU-2, CUM-2.
(4) Located: none. Listed (as "Augusta Mie"): CvR-4.

AURETII
(1) "*Rose Auretii.* Lith. & col[d.] by Amana Society. Amana, Iowa County, Iowa." (Printed from stone in AMAH collection.) Located: (uncolored), AMAH; (colored), AMAH.
(2) Located: none. Listed: CvR-4.

BALTIMORE BELLE
Located: none. Listed (as "Baltimore Belle"): CvR-4.

BARON PREVOST. (Incorrect for "BARONNE PREVOST," see below.)

BARONNE PREVOST
(1) "*Baronne Prevost.*" (Small plate, 27 × 17 cm. Caption written in ink. Same design, with only minor changes in the design and coloring, as the "Pius the 9th" and "Caroline de Sansel" plates.) Located: CvR-1.
(2) Located: none. Listed: CU-2, CUM-2.
(3) "*Baron Prevost.*" (A variation of the"Baronne Prevost" design, #1.) Located: APC.
(4) "*Baron Prevost.*" Located: none. Listed: CvR-4.
(5) "*Baron Brevost* [*sic*]." (Uncolored.) Located: AMAH.

CAROLINE DE SANSAL
(1) "*Caroline De Sansel.*" (Small plate, 27 × 17 cm. Caption written in ink. Same design as the "Pius the 9th" plate except for differently colored blossom.) Located: CvR-1.
(2) "*Caroline De Sansel.*" (Simplified version of #1. Same design and lettering as the Joseph Prestele plate in the *Horticulturist*, vol. 4, n.s. [November 1854].) Located: CvR-2.
(3) "*Caroline De Sansel.*" (Large plate 32 × 24.5 cm.) Located: RU-1.
(4) Located: none. Listed: CU-2, CUM-2.
(5) Located: none. Listed (as "Carolina de Sansal"): CvR-4.

COTTAGE ROSE
Located: none. Listed: CvR-4.

CRESTED MOSS (See MOSS, Crested.)

DEVONIENSIS
Located: none. Listed (as "Tea Rose Devoniensis"): CvR-4.

GEANT DES BATAILLES
(1) "*The Rose Geant Des Batailles* (Giant of Battles)" (Adapted from the Joseph Prestele plate in the *Horticulturist*, vol. 4, n.s. [December 1854].) A theorem and freehand version of this design was published ca. 1859 with the imprint "D. M. Dewey's Series of Fruits, Flowers and Ornamental Trees. Rochester, N.Y." Located: CvR-2, CvR-3.
(2) (Same as #1?) Located: RU-1.
(3) (Same as #1?) Located: none. Listed: CU-2, CUM-2.
(4) Located: none. Listed: CvR-4.

GEM OF THE PRAIRIE
Located: none. Listed (as above): CvR-4.

GENERAL JACQUEMINOT
(1) "*Gen. Jacqueminot Rose.*" Located: RU-2.
(2) Located: none. Listed (as "General Jaquimenot"): CvR-4.

GENERAL WASHINGTON (See THE WASHINGTON ROSE.)
Located: none. Listed (as "General Washington"): CvR-4.

GEORGE THE FOURTH
Located: none. Listed (as above): CvR-4.

GLOIRE DE DIJON
Located: none. Listed (as above): CvR-4.

GLORY OF MOSSES
Located: none. Listed (as above): CvR-4.

HERMOSA
"*The Hermosa Rose.* Lith. by Amana Society. Homestead, Iowa." (Uncolored) Located: AJM.

JOHN HOPPER
Located: none. Listed (as above): CvR-4.

KING OF THE PRAIRIE
Located: none. Listed (as above): CvR-4.

LA REINE
(1) "*La Reine.*" (By E. B. & E. C. Kellogg, Hartford, Connecticut, ca. 1859. Imprint removed by trimming.) Located: CvR-3.
(2) "*La Reine.* Lith. & col[d.] by Amana Society, Amana, Iowa County, Iowa." (Same design as #1?) Located: APC.
(3) Located: none. Listed: CU-2, CUM-2.
(4) Located: none. Listed: CvR-4.

LION DES COMBATS
(1) Located: none. Listed (as "Leon des Combats"): CU-2, CUM-2.
(2) Located: none. Listed (as "Leon Des Combats"): CvR-4.

LORD RAGLAN
(1) Located: none. Listed: CU-2, CUM-2.
(2) Located: none. Listed: CvR-4.

MADAME PLANTIER
(1) "*Madame Plantier Rose.*" Located: APC.
(2) "*Madame Plantier Rose.*" (Same as #1?) Located: CvR-1.
(3) Located: none. Listed: CvR-4.

MADAME DE ROUGEMONT
Located: none. Listed (as "Madam de Rougemont"): CvR-4.

MARECHAL NIEL
Located: none. Listed (as "Marshall Neil"): CvR-4.

MOSS
(1) "*Moss Rose.*" Located: CvR-2.
(2) "*Moss.*" (With imprint, "D. M. Dewey, Rochester, N.Y.," ca. 1859. Printed from a Prestele impression, or a Dewey copy of a Prestele design.) Located: CvR-3.
(3) "*Moss Rose.*" (Printed from stone in AMAH collection.) Located: AMAH.

MOSS ROSE. Crested
Located: none. Listed: CvR-4.

MOSS ROSE. Perpetual White
Located: none. Listed (as above): CvR-4.

MOSS ROSE. Red
Located: none. Listed (as above): CvR-4.

MOSS ROSE. Salet
Located: none. Listed (as above): CvR-4.

PANACHÉE D'ORLEANS
(1) "*Panache D'Orleans.*" Located: CvR-2. (A theorem and freehand version of this plate was published as "Panach D'Orleans," with the imprint, "D. M. Dewey's Series of Fruits, Flowers, and Ornamental Trees, Rochester, N.Y.," ca. 1859. Located: CvR-2.)
(2) "*Panache D'Orleans.*" Located: RU-1.
(3) (Same as #2?) Located: none. Listed: CU-2, CUM-2.
(4) Located: none. Listed (as "Panach de Orleans"): CvR-4.

PERSIAN YELLOW (See "Yellow Persian," a synonym.)
(1) "*Persian Yellow Rose.*" Located: CvR-2.
(2) "*Yellow Persian.*" Located: none. Listed: CU-2, CUM-2.

PIUS IX
(1) "*Pius The 9th.*" (Caption written in ink. Small plate, 27 × 17 cm. Same design as the "Caroline de Sansal" plate except for a differently colored blossom.) Located: CvR-1.
(2) Located: none. Listed (as "Pius The Ninth"): CvR-4.

PRINCE ALBERT
Located: none. Listed (as above): CvR-4.

QUEEN OF THE PRAIRIES
(1) "*Queen of the Prairies.* Lith. & col[d.] By Wm. H. Prestele. Artist. Iowa City, Iowa." Located: APC.
(2) (Same as #1?) Located: none. Listed: CU-2, CUM-2.
(3) "*The Queen of the Prairie Rose.*" Located: (uncolored), AMAH; (colored), APC.

SEVEN SISTERS
"Seven Sisters/Lith. & col[d.] by W. H. Prestele, Iowa City, Iowa." Located: AES.

"SINGLE PURPOSE ROSE. ROSA GALLICA"
"*Single Purpose Rose. Rosa Gallica.* Drawn from nature & colored by G. Prestele." (Watercolor; caption written in ink.) Located: APC.

SOLFATERRE (See AUGUSTA.)

SOUVENIR DE LA MALMAISON.
(1) "*The Souvenir De La Malmaison Rose.*" (Uncolored). Located: AJM. (A simpler version of this design, attributed to Joseph Prestele, appeared as an illustration in the *Horticulturist*, vol. 3, n.s. [August 1848]. That illustration, with minor changes, was copied in a theorem and freehand version by D. M. Dewey and issued in 1859 with his imprint. The Prestele nurserymen's plate listed here is almost identical to Dewey's in design.)
(2) Located: none. Listed: CvR-4.

WASHINGTON (See GENERAL WASHINGTON.)
"*The Washington Rose.*" Located: AES.

TREE ROSE
"*Weeping Tree Rose*, with 15 Roses in bloom, of different colors, a superb fine plate." Located: none. Listed. CvR-4.

YELLOW PERSIAN (See PERSIAN YELLOW ROSE, a synonym.)

Ornamental Trees

"AMERICAN ARBOR VITAE"
Located: none. Listed: CvR-4.

"AMERICAN ARBOR VITAE HEDGE"
Located: none. Listed: CvR-4.

"AMERICAN LARCH"
Located: none. Listed: CvR-4.

"AMERICAN PINE"
Located: none. Listed: CvR-4.

"BALSAM FIR"
Located: none. Listed: CvR-4.

"CAMPERDOWN WEEPING ELM"
Located: none. Listed: CvR-4.

"CHIONANTHUS VIRGINICA. White Fringe Tree"
(1) As above. Located: AJM.
(2) Located: none. Listed (as "White Fringe"): CvR-4.

"DOUBLE WHITE FLOWERING ALMOND. DOUBLE CRIMSON FLOWERING PEACH."
A. (With blossoming specimens shown separated.)
(1) Caption as above. (Lithographed with penciled details.) Located: CvR-2. The same design, by Joseph Prestele, Sr., was used as a frontispiece in both the colored and uncolored editions of the *Horticulturist*, n.s., vol. 5, no. 5 (May 1855). A theorem and freehand copy of that illustration was issued ca. 1859 with the imprint of "D. M. Dewey, Rochester, N.Y." Listed: CvR-3.
(2) (Same design as #1. Uncolored. Printed from the stone at AMAH.) Located: AMAH.
(3) (Same design as #1?) Located: none. Listed: CvR-4.
B. (With blossoming specimens shown entwined.) Caption as above. Colored and uncolored plates. Located: AMAH.

"ENGLISH JUNIPER"
Located: none. Listed: CvR-4.

"EUROPEAN LARCH"
Located: none. Listed: CvR-4.

"HORSE CHESTNUT"
Located: none. Listed: CvR-4.

"HONEY LOCUST HEDGE"
Located: none. Listed: CvR-4.

"IRISH JUNIPER"
Located: none. Listed: CvR-4.

KILMARNOCK WEEPING WILLOW
"*Kilmarnock Weeping*"/Lith. & col[d.] by W. H. Prestele, Iowa City, Iowa." Located: AJM.

"MAGNOLIA"
(Uncolored.) Located: AMAH. (Apparently Gottlieb Prestele's file copy with penciled notations that (25?) copies were printed December 2, 1864; 100, December 6, 1864; 50, February 16, 1864; 50, February 11, 1867; 50, March 14, 1871; 100, June 28, 1872. On February 26, 1873, he noted that he removed the image from the stone.)

"MAGNOLIA GLAUCA" (*M. virginiana*)
Located: none. Listed: CvR-4.

"MAGNOLIA SOULANGIANA"
Located: none. Listed: CvR-4.

"MAGNOLIA. Umbrella Tree." (*M. tripetala*)
(1) Located: AMAH.
(2) Located: none. Listed (as "Magnolia Umbrella Tree"): CvR-4.

"MOUNTAIN ASH"
Located: none. Listed: CvR-4.

"NORWAY SPRUCE"
Located: none. Listed: CvR-4.

"OSAGE ORANGE HEDGE"
Located: none. Listed: CvR-4.

"ROSEMARY LEAVED WILLOW"
Located: none. Listed: CvR-4.

Shrubs

"AFRICAN TAMARIX" (See "TAMARIX AFRICANS")

"ALTHEA, or ROSE OF SHARON"
"*Althea or Rose of Sharon*/Lith & col[d.] by W. H. Prestele Iowa City Iowa." Located: AES.

"CALYCANTHUS"
Located: none. Listed: CvR-4.

"DAPHNE MEZEREUM"
Located: none. Listed: CvR-4.

"DEUTZIA CANESCENS"
Located: none. Listed: CU-2.

"DEUTZIA CRENATA, fl. pl."
Located: none. Listed: CvR-4.

"DEUTZIA GRACILIS"
Located: none. Listed: CvR-4.

"DEUTZIA SCABRA"
(1) "*Deutzia Scabra*/Rough leaved Deutzia." (Image for this plate is on a stone at AMAH.) Located: none.
(2) (Same as #1?) Located: none. Listed: CU-2, CUM-2.
(3) Located: none. Listed: CvR-4.

"FORSYTHIA VIRIDISIMA, or GOLDEN BELL"
Located: none. Listed: CvR-4.

"GENISTA SCOPARIS [*CYTISUS scoparius*]/Scotch broom"
Located: none. Not listed on CU-2, or CUM-2. Image for this plate on stone at AMAH.

LILAC, Purple and White
"*Lilac Purple & White.*/Lith. & col[d.] by W. H.

Prestele, Iowa City, Iowa." Located: AMAH.

"LONICERA (3 specimens)"
(1) Located: none. Listed (as above): CU-2, CUM-2.
(2) Located: none. Listed (as "Lonicera, or Honeysuckle"): CvR-4.

"LONICERA TARTARICA"
(1) "*Lonicera Tartarica*/Tartarian Honeysuckle." (The image for this plate is on a stone at AMAH.)
(2) (Same design as #1?) Located: none. Listed: CU-2, CUM-2.
(3) Located: none. Listed (as "Lonicera Tartarica, or Upright Honeysuckle"): CvR-4.

"MAHONIA AQUIFOLIA" (*M. aquifolium*)
Located: none. Not listed on CU-2, or CUM-2. (The image for this plate is on a stone at AMAH.)

"PURPLE FRINGE" (*Cotinus coggygria atropurpurea*)
Located: none. Listed: CvR-4.

"PYRUS JAPONICA" (*Chaenomeles japonica*)
(1) Located: CvR-2.
(2) Located: none. Listed: CU-2, CUM-2.
(3) Located: none. Listed (as "Pyrus Japonica, or Japan Quince"): CvR-4.

"RHODODENDRON, 3 diff. varieties, including variegated"
Located: none. Listed (as above): CvR-4.

"RIBES BEATONI, or Flowering Currants" (*R. odoratum*?)
Located: none. Listed: CvR-4.

"ROSE ACACIA, OR MOSS LOCUST" (*Robinia hispida*)
Located: none. Listed: CvR-4.

"SNOW BALL" (See VIBURNUM OPULUS. Snow Ball.)
(1) Located: none. Listed: CUM-2, CU-2.
(2) Located: none. Listed: CvR-4.

SIPIREA BILLARDI" (*Spiraea billiardi*)
Located: none. Listed: CvR-4.

"SPIREA./I. Billardi [*Billiardi*]. II. Salicifolia."
(1) Located: CvR-1.

"SPIREA CALLOSA" (*Spiraea japonica*)
(1) Located: APC.
(2) Located: none. Listed: CvR-4.

"SPIREA OPULIFOLIA" (*Spiraea ovalifolia*?)
(1) Located: none. Listed: CvR-4.
(2) Located: none. Listed: CU-2, CUM-2.

"SPIREA PRUNIFOLIA" (Bridalwreath Spirea)
Located: none. Listed: CvR-4.

"SPIREA/I Prunifolia fl. pl./I Double flowering Plum leaved./II Lanceolata./II Lance leaved Spirea." (As above.) Located: AMAH. Listed: CU-2, CUM-2, AAS-3. (The image for this plate is on a stone at AMAH.)

"SPIREA REVISII" (Reeves S. *S. catoniensis.*)
Located: none. Listed: CvR-4.

"SPIREA SPIREA"
Located: none. Listed: CvR-4.

"SPIREA ULMARIA" (*Filipendula ulmaria*)
Located: none. Listed: CvR-4.

"SPIREA ULMARIA ALBA" (*Filipendula purpurea alba*)
Located: none. Listed: CU-2, CUM-2.

"SPIREA ULMARIA ROSEA" (*Filipendula rubra*)
Located: none. Listed: CU-2, CUM-2.

"SYRINGA, or Mock Orange."
"*Syringa or Mockorange*/Lith. & Col[d.] by W. H. Prestele, Iowa City, Iowa." Located: AES.

TAMARIX AFRICANA/African Tamarix"
(1) Located (colored and uncolored): AMAH. (The stone used for printing this plate is preserved at AMAH.)
(2) Located: none. Listed: CvR-4.

"VIBURNUM LANTANA./Wayfaring Tree"
(1) Located: AES.
(2) (Same design as #1?) Listed: CU-2, CUM-2, AAS-3.

"VIBURNUM OPULUS./SNOW BALL."
Located: none. Listed (as "Snow Ball"): CU-2, CUM-2, AAS-3. (The stone used for printing this plate is preserved at AMAH.)

"WIEGELIA ROSEA" (*Weigela florida*)
(1) "*Wiegelia Rosea*/Rose—?—." (The stone used for printing this plate is preserved at AMAH.)
(2) (Same design as #1?) Located: none. Listed: CU-2, CUM-2.
(3) Located: none. Listed: CvR-4.

"WIEGELIA VARIEGATED LEAVED" (*W. florida variegata*)
Located: none. Listed: CvR-4.

Vines

"BIGNONIA RADICANS/Trumpet Flower" (*Campsis radicans*)
(1) Located: APC, CvR-2.
(2) Located: none. Listed (as "Bignonia Radicans or Trumpet Flower"): CvR-4.
(3) "*Bignonia Radicans/ Trumpet Flower.*" (Poor color, caption stamped on plate in peacock blue ink.) Located: AES.

"CLEMATIS JACKMAN" (*C. jackmani*)
Located: none. Listed: CvR-4.

"LONICERA/I Peryclimenum [*Periclymenum*]. Woodbine II Sempervirens. Monthly Honeysuckle. III Frazerii [*Flava*?] Strawcolored Honeysuckle."
(1) Located: AMAH.
(2) (Same design as #1?) Same caption except for abbreviations; "Strawcol[d.]" and "Honeyskle.") Located: CvR-5. Listed (as "Lonicera, or Honeysuckle"): CvR-4.

"WISTARIA PURPLE" (*Wistaria simensis*)
Located: none. Listed: CvR-4.

"WISTARIA WHITE" (*Wistaria simensis alba*)
Located: none. Listed: CvR-4.

Index

The Smithsonian Institution gratefully acknowledges the generous support of these contributors:

THE BARRA FOUNDATION, INC.
MRS. ALFRED E. BISSELL
MR. AND MRS. GEORGE P. BISSELL, JR.
MR. AND MRS. LAMMOT DUPONT COPELAND
MRS. HENRY B. DUPONT
MRS. NICHOLAS R. DU PONT
MRS. ELWYN EVANS
MR. AND MRS. PAUL MELLON
STANLEY AND POLLY STONE FOUNDATION

This book was produced by the Smithsonian Institution Press
Printed by Balding + Mansell Printers, Wisbech, Cambs
Set in Bembo by Graphic Composition, Inc., Athens, Georgia
Edited by Jeanne Sexton
Designed by Carol Hare Beehler